Hands-On Herpetology

Exploring Ecology and Conservation

Dedicated to
Whit Gibbons:
Scholar, Mentor, and Friend

NATIONAL SCIENCE TEACHERS ASSOCIATION
ARLINGTON, VIRGINIA

Hands-On Herpetology

Exploring Ecology and Conservation

Rebecca L. Schneider   Marianne E. Krasny   Stephen J. Morreale

Illustrations by Tami Tolpa

ARLINGTON, VIRGINIA

NSTA Press, NSTA Journals, and the NSTA website deliver high-quality resources for science educators.

Featuring *sci*LINKS®—a new way of connecting text and the Internet. Up-to-the-minute online content, classroom ideas, and other materials are just a click away. Go to page xv to learn more about this new educational resource.

On the Cover: Design and illustrations of green treefrog, eastern box turtle, and eastern newt by Linda Olliver.

Hands-On Herpetology: Exploring Ecology and Conservation
NSTA Stock Number: PB163X
ISBN: 0-87355-197-4
Library of Congress Control Number: 200193729
Printed in the USA by Banta Book Group–Harrisonburg
Printed on recycled paper

The mission of the National Science Teachers Association is to promote excellence and innovation in science teaching and learning for all.

NSTA Press
1840 Wilson Boulevard
Arlington, Virginia 22201-3000
www.nsta.org

Table of Contents

About the Authors vii
Preface: Why an Activity Guide Focused on Herps? ix
Acknowledgments x
Introduction xi
How to Use *Hands-On Herpetology* xi
The *National Science Education Standards* xii
*sci*LINKS xv

Section I. Getting Started 1

Care and Handling of Live Herps 2
Safety Concerns 5
Permits and Regulations 7
Basic Facts about Herps 7

Section II. Biology: Up Close and In Hand 13

Chapter 1. Morphology—Getting into Shape(s) 15
Activity 1.1 Herps Inside-Out 17
Chapter 2. Skin—It's Packaging Plus 21
Activity 2.1 Skin Textures 22
Activity 2.2 Lurking Lizards 23
Chapter 3. Defense Systems—On Guard! 25
Activity 3.1 Mimics Survive 27
Activity 3.2 Don't Mess with Me 27
Chapter 4. Multiplying Herps—Reproduction and Development 31
Activity 4.1 Amphibian Life Cycles 34
Chapter 5. Breeding Behavior 37
Activity 5.1 Visit an Evening Chorus 38
Chapter 6. Life Span and Life History 41
Activity 6.1 Long-Lived Turtles 42
Activity 6.2 Estimating Madness 43
Activity 6.3 Turtle Shell and Egg Hunt 46
Chapter 7. Herps through the Ages—Evolution and Extinction 49
Activity 7.1 Dinosaur Detectives 51

Section III. Herps: A Role in the Bigger Ecosystem 55

Chapter 8. Monitoring Species Diversity and Abundance 57
Activity 8.1 Herp Checklist 59
Chapter 9. Herps and Their Homes 65

Activity 9.1 Herp Walk 67
Chapter 10. Food Webs 69
Activity 10.1 Discover a Pond Food Web 71
Chapter 11. Upland-Aquatic Linkages 73
Activity 11.1 Sampling Migratory Herps with a Temporary Drift Fence 74

Section IV. Conservation and Management 77

Chapter 12. Habitat Fragmentation 79
Activity 12.1 Herp Crossing Alert! 80
Chapter 13. Wetland Loss and Restoration 83
Activity 13.1 Frog-Friendly Ponds 84
Chapter 14. Invisible Threats 87
Activity 14.1 Threat Assessment 88
Chapter 15. Species Conservation 91
Activity 15.1 Recycling Unwanted Herp Pets 92
Chapter 16. Crisis Intervention 95
Activity 16.1 Seaturtle Head Start and Rescue Programs 98

Section V. Herp Malformations and Declines: A Scientific Inquiry 99

Chapter 17. Mystery of the Malformed Frogs 101
Activity 17.1 The Students' Discovery 103
Chapter 18. Amphibian Population Declines 105
Activity 18.1 Why Care about Declines and Malformations? 107
Chapter 19. Becoming Citizen-Scientists: Surveying Amphibian Malformations and Declines 109
Activity 19.1 Monitoring Amphibian Malformations 111
Activity 19.2 Surveying Amphibian Populations 112
Chapter 20. Solving the Mystery of Amphibian Malformations 113
Activity 20.1 Environmental Research 114
Chapter 21. Beyond Research: Scientific Disagreement, Ethics, and Policy 123
Activity 21.1 Scientific Disagreement 124
Activity 21.2 Science and Policy 127
Activity 21.3 Ethics and Policy 127
Activity 21.4 Herp Websites 128

Appendix: Herp Species Accounts 133

About the Authors

Rebecca L. Schneider, Ph.D., is an ecologist studying wetland and aquatic ecosystems. In her research, she examines the interactions between plant communities and flowing water, both aboveground and belowground. Primary topics in her research include seed dispersal, hydrologic transport of nutrients, filtration of contaminants by plants, and streamside buffering. As an assistant professor in the Department of Natural Resources at Cornell University, Rebecca teaches about sustainable management of water resources and, through her outreach programs, provides guidance to local communities and state and national government agencies.

Marianne E. Krasny, Ph.D., is an associate professor, department extension leader, and director of graduate studies in the Department of Natural Resources at Cornell University. Through her work with the Cornell Environmental Inquiry Program, Marianne develops and evaluates university outreach programs designed to engage students and teachers in authentic environmental sciences research.

Stephen J. Morreale, Ph.D., is a zoologist whose research is focused in the field of herpetology. He has worked for many years studying turtle ecology and is a world-renowned expert on seaturtles. As a senior research associate in the Department of Natural Resources at Cornell University and the associate director of research at Cornell's Arnot Teaching and Research Forest, he devotes considerable time to interacting with students, conducting research, and teaching conservation ecology. Stephen also is active in the conservation community, working with several national agencies and international organizations.

Preface

Why an Activity Guide Focused on Herps?

"Herps" are all around us. They include amphibians, such as frogs, toads, and salamanders, and reptiles, such as snakes, lizards, and turtles. Yet surprisingly, herps have been largely ignored as a tool for educating youth about biology, ecology, and conservation. Thus, this educator's guide addresses a strong need for educational resources focused on the science and conservation of amphibians and reptiles.

Because *herps are plentiful, diverse, and nearby*, they offer many possibilities for teaching science and conservation to young people. Unlike birds or mammals, most frogs, salamanders, and turtles are relatively docile and easy to hold. When treated properly, they do not show fear or threatening postures, which can be a distraction to students and an obstacle to learning. Also, amphibians generally are harmless, and in much of the northern United States and eastern Canada there are few poisonous species of reptiles, so many fears and concerns can be set aside.

Young people and adults studying herps will find fascinating differences among the various groups in their basic physiology, morphology, behavior, and adaptations to their environments. Thus, *herps are excellent organisms to demonstrate concepts about biology and ecology*, ranging from camouflage and thermoregulation, to food webs and competition.

Perhaps the most important reason for developing an educator's guide focusing on herps is the *urgent need for expanding conservation efforts* directed at these organisms. Because herps usually are hidden from sight, we often are not aware of their amazing abundance and importance to the overall environment. However, herps are an integral part of many ecosystems, and serve essential roles as predators on small animals as well as food for larger ones. At the same time, herp populations worldwide are suffering severe declines, and many species have become extinct in recent years. Additionally, some populations of herps in North America have experienced alarmingly high rates of malformations in legs, tails, eyes, and internal organs. Scientists and the public are still searching for explanations for these mysterious declines and malformations—with the objective of preserving the diversity and richness of amphibians and reptiles.

We hope that you and the young people with whom you work will enjoy learning about the fascinating world of frogs, toads, salamanders, snakes, and turtles. At the same time, we hope you will join us in efforts to conserve these amazing and important creatures.

Acknowledgments

We appreciate the contributions of Martin Schlaepfer to this project, including the species accounts. We thank Margaret Corbit for assistance during the development phase. We also are grateful for the enthusiastic critiques provided by educators who pilot-tested many of the activities in this guide, including Andy Turner, Sanford Smith, Bonnie Peck, Keith Koupel, Carolyn Klass, Susan and Paul Grimes, Kent Gaerthe, Kimberly Fleming, and Gwen Curtis. Many others provided insights, suggestions, and support that have improved the book immensely, including Kraig Adler, Al Breisch, Nancy Bowers, Howard Evans, Harry Greene, Gretchen Finley, John Maerz, Karen Poiani, Mike Richmond, and Kelly Zamudio. Funding to initiate this project was provided as a grant from the U.S. Fish and Wildlife Service.

The book's reviewers were Lisa Robinson, a biology teacher at Oxford High School, Oxford, Alabama, and Juli Werth, a teacher of sixth-grade integrated science at Riverview Middle School, Huntington, Indiana. Kenneth Roy, K-12 director of science and safety for the Glastonbury, Connecticut, Public Schools and chairperson of the NSTA Science Safety Advisory Board, reviewed the section on safety.

All the illustrations are by Tami Tolpa, San Francisco, California, except those noted below.

The NSTA project editor for the book was Judy Cusick. Linda Olliver designed the book and the cover, did book layout, and drew the illustrations on pages 18, 19, 45, 46, and 53. Catherine Lorrain-Hale coordinated production and printing.

Introduction

How to Use *Hands-On Herpetology*

The goal of *Hands-On Herpetology* is to provide an introduction to the study of reptiles and amphibians and to present opportunities for young people ages 10–18 (grades 5–12) to become involved in their conservation. This book is designed as an instructional guide for many educators, including upper-elementary, middle, and high school teachers, 4-H and summer youth camp instructors, and nature center educators. The material provides a thorough introduction for newcomers to the world of amphibians and reptiles, and also presents relevant and challenging activities for more experienced herpetologists.

The information and activities have been divided into five sections. In Section I, we explain the basics of interacting with herps, including handling, safety, permits, and regulations. We also provide an introduction to herp biology. This section provides a critical underpinning for the remainder of the book. The next three sections include background information and hands-on activities to illustrate basic principles that are common to many species and their communities: We focus on the biology of reptiles and amphibians (Section II), their ecology (Section III), and their conservation (Section IV).

In Section V, students examine two critical conservation issues currently affecting amphibians—malformation and declining populations. In doing so, they become familiar with many issues scientists face when researching environmental phenomena. For example, students learn about different research approaches, ranging from monitoring to controlled experiments, and about how policymakers use research findings. They also learn how careful observations conducted by young people and volunteers, as well as by professional scientists, can contribute to solving disturbing environmental problems. In addition, they learn about opportunities to use the Internet to link up with groups of students, volunteers, and scientists who are monitoring herp populations and sharing their data. Finally students see how the knowledge and skills they learned through the activities in previous sections can be applied to solve the double mysteries of herp decline and amphibian malformations.

There is considerable variety among the 29 activities presented. Although most of the activities can be conducted anywhere in North America, we draw mainly on examples of animals from the northeastern states. About half of the activities take place indoors, and the other half outdoors. For the outdoor activities, some are best conducted in early spring whereas others work well throughout the warmer months when herps are active. The activities also vary in their degree of difficulty; some are most suitable for upper-elementary and middle school students, whereas others may be appropriate for high school students. You will need to consider which activities are most appropriate given your students' abilities, the resources available, and season of the year.

Hands-On Herpetology is not intended to be a comprehensive, sole source reference. Instead we strongly recommend that educators complement its use with field guides identifying amphibians and reptiles in their own regions. As an appendix, we have included example species accounts (pages 133–145) for three amphibians (the spotted salamander, the American toad, and the eastern newt) and one reptile (the snapping turtle). Such accounts can help you identify herp species and familiarize you with their natural history and conservation status. You may want to contact your state natural resources agency or a local university for more accounts of species that occur in your region.

The *National Science Education Standards*

In 1996 the National Research Council of the National Academy of Sciences published the *National Science Education Standards*, a set of guidelines for the teaching of science in grades K–12. These standards move science teaching away from vocabulary lists and "canned" laboratories with pre-set outcomes toward the teaching of science as an active process that requires not only knowledge but also reasoning and thinking skills.

The activities in this book are designed to embrace the philosophy set forward by the creators of the National Standards. As students carry out the activities, they will take an active approach to learning through hands-on activities, group discussions, and participation in Internet-facilitated research activities. In general, Section II of the guide focuses on Life Science standards; Section III focuses on Life Science and Science as Inquiry; Section IV focuses on Science in Personal and Social Perspectives; and Section V focuses on Science as Inquiry, Science in Personal and Social Perspectives, and the History and Nature of Science. The *National Science Education Standards* for grades 5–8 and grades 9–12 covered in each chapter are indicated at the beginning of the chapter. In addition, two Standards matrixes follow—"Correlations with the *National Science Education Standards* for Grades 5–8" and "Correlations with the *National Science Education Standards* for Grades 9–12."

Figure i.1—Correlations with the *National Science Education Standards* for Grades 5–8

		Section II						Section III					Section IV					Section V				
Content Standard	Topic	Ch. 1	Ch. 2	Ch. 3	Ch. 4	Ch. 5	Ch. 6	Ch. 7	Ch. 8	Ch. 9	Ch. 10	Ch. 11	Ch. 12	Ch. 13	Ch. 14	Ch. 15	Ch. 16	Ch. 17	Ch. 18	Ch. 19	Ch. 20	Ch. 21
(A) Science as Inquiry	Abilities necessary to do scientific inquiry						•		•	•	•	•						•	•	•	•	•
	Understanding about scientific inquiry																	•	•	•	•	•
(C) Life Science	Structure and function in living systems	•	•	•																		
	Regulation and behavior			•								•										
	Reproduction and heredity				•	•																
	Populations and ecosystems						•				•			•								
	Diversity and adaptations of organisms							•	•	•												
(F) Science in Personal and Social Perspectives	Populations, resources, and environments												•	•	•	•	•					
	Natural hazards												•	•	•	•	•	•	•	•	•	•
(G) History and Nature of Science	Science as human endeavor																	•	•	•	•	•
	Nature of scientific knowledge																		•	•	•	•

Figure i.2—Correlations with the *National Science Education Standards* for Grades 9–12

		Section II						Section III					Section IV					Section V				
Content Standard	**Topic**	**Ch. 1**	**Ch. 2**	**Ch. 3**	**Ch. 4**	**Ch. 5**	**Ch. 6**	**Ch. 7**	**Ch. 8**	**Ch. 9**	**Ch. 10**	**Ch. 11**	**Ch. 12**	**Ch. 13**	**Ch. 14**	**Ch. 15**	**Ch. 16**	**Ch. 17**	**Ch. 18**	**Ch. 19**	**Ch. 20**	**Ch. 21**
(A) Science as Inquiry	Abilities necessary to do scientific inquiry						•		•	•	•	•						•	•	•	•	•
	Understanding about scientific inquiry																	•	•	•	•	•
(C) Life Science	Behavior of organisms			•	•	•						•										
	Biological evolution							•														
	Interdependence of organisms										•											
(F) Science in Personal and Social Perspectives	Population growth						•															
	Natural and human-induced hazards							•						•	•	•	•	•	•	•	•	•
	Natural resources												•									
	Environmental quality												•	•	•	•	•	•	•	•	•	•
	Science and technology in local, national, and global challenges												•									
(G) History and Nature of Science	Science as human endeavor																	•	•	•	•	•
	Nature of scientific knowledge																	•	•	•	•	•

In *Hands-On Herpetology,* NSTA brings you *sci*LINKS, a new project that blends the two main delivery systems for curriculum—books and telecommunications—into a dynamic new educational tool for children, their parents, and their teachers. *sci*LINKS links specific science content with instructionally rich Internet resources. *sci*LINKS represents an enormous opportunity to create new pathways for learners, new opportunities for professional growth among teachers, and new modes of engagement for parents.

In this *sci*LINKed text, you will find an icon near several of the concepts you are studying. Under it, you will find the *sci*LINKS URL (*www.scilinks.org*) and a code. Go to the *sci*LINKS website, sign in, type the code from your text, and you will receive a list of URLs that are selected by science educators. Sites are chosen for accurate and age-appropriate content and good pedagogy. The underlying database changes constantly, eliminating dead or revised sites or simply replacing them with better selections. The ink may dry on the page, but the science it describes will always be fresh. *sci*LINKS also ensures that the online content teachers count on remains available for the life of this text. The *sci*LINKS search team regularly reviews the materials to which this text points—revising the URLs as needed or replacing web pages that have disappeared with new pages. When you send your students to *sci*LINKS to use a code from this text, you can always count on good content being available.

The selection process involves four review stages:

1. First, a cadre of undergraduate science education majors searches the World Wide Web for interesting science resources. The undergraduates submit about 500 sites a week for consideration.
2. Next, packets of these webpages are organized and sent to teacher-webwatchers with expertise in given fields and grade levels. The teacher-Webwatchers can also submit webpages that they have found on their own. The teachers pick the jewels from this selection and correlate them to the National Science Education Standards. These pages are submitted to the *sci*LINKS database.
3. Scientists review these correlated sites for accuracy.
4. NSTA staff approve the webpages and edit the information provided for accuracy and consistent style.

*sci*LINKS is a free service for textbook and supplemental resource users, but obviously someone must pay for it. Participating publishers pay a fee to NSTA for each book that contains *sci*LINKS. The program is also supported by a grant from the National Aeronautics and Space Administration (NASA).

Section I

Getting Started

Before starting any of the activities in this guide, we strongly recommend that you read the information in this section. Here we cover the critically important issues of how to handle amphibians and reptiles, and safety issues concerning these animals and field activities in general. We also discuss permits and regulations relating to the collecting, handling, and raising of herps. At the end of this section, you will find an introduction to the biology of amphibians and reptiles—"Basic Facts about Herps"—which provides important background for understanding the subsequent chapters.

- Care and Handling of Live Herps
- Safety Concerns
- Permits and Regulations
- Basic Facts about Herps

Care and Handling of Live Herps

When handling any live animal, it is important to always keep two safety issues in mind: first is the safety of the person who is searching for or holding the animal, and second is the safety and welfare of the animal itself.

People may believe naively that because herps do not show facial expressions or fear, or have obvious defensive postures, they cannot be hurt. However, herps are still very much living organisms, vulnerable to stress and pain caused by improper handling and confinement. Humans also can harm entire herp populations by destroying their homes or nests, as well as by disturbing their habitats on a broader scale. If you follow the guidelines below for safe capture, care, and handling of herps, you will go a long way toward ensuring both your safety and that of the reptiles and amphibians you are studying.

When searching for herps, *take care not to destroy their homes or injure the animals*. Lift rocks gently and lower them carefully back to their original position, so as not to crush soft-bodied salamanders and other creatures living underneath. Return logs with the original, moist side down because they already may be decomposing and may provide homes for many organisms. Don't tear up logs or tree stumps to get to salamanders, lizards, or snakes because this permanently destroys their nests and their homes.

Keep handling of herps to a minimum in order to protect the animals. Amphibians have delicate skin that needs to be kept moist. Continuous handling dries out their skin and removes the mucous-like protective covering that is present on many amphibians. Without this covering, the amphibian's skin may suffer from abrasion and infection. Even

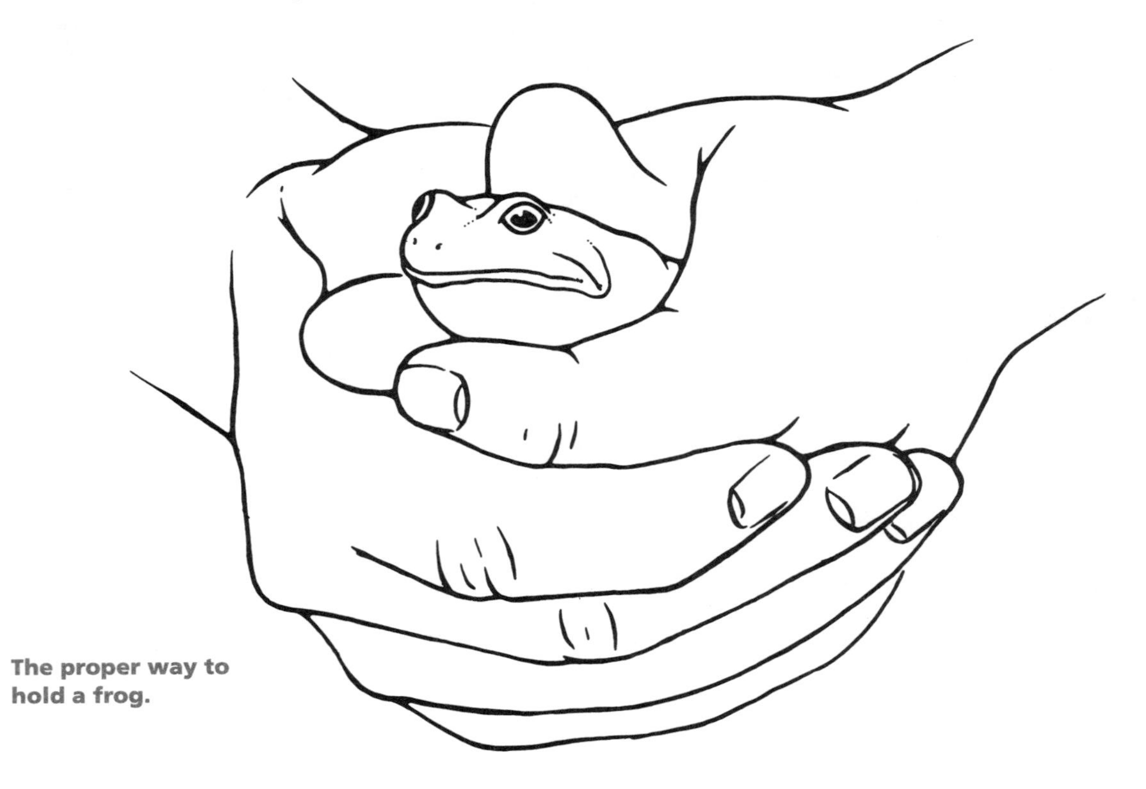

The proper way to hold a frog.

tough-skinned reptiles can be stressed by handling and restraint. Therefore, the less they are subjected to handling, the better. A useful method for looking closely at and displaying animals to a large group is to place the animal in a clear, plastic container. With the addition of a little moisture from leaves or a wet towel, a clear container can make a very suitable viewing device.

If an animal feels secure and comfortable, it is less likely to try to escape or bite. Therefore *hold all organisms gently, but firmly*. Salamanders and frogs have particularly tiny and fragile limbs, toes, and tails. You can hold them by gently cupping your hand to support their bodies. To look at them more closely, you may have to restrain their limbs and keep them from wiggling without squeezing too hard. This takes a little practice, and adults should assist children in holding the animals until the children are comfortable and competent at it. A great technique for the newcomer and experienced alike is to always crouch or sit on the ground when restraining an animal. That way, when they squirm out, there won't be any shock or damage from crashing to the ground.

Snakes need to have their bodies supported while their heads are being immobilized. Use the thumb and fingers to gently restrain the sides of their heads near the neck, but don't let the body flail about freely. It is best to just observe a snake from a bit of a distance if the species is unknown. Whether venomous or not, snakes of many species will bite to protect themselves. Also, don't pick up snakes that have obviously just eaten. This will be apparent from a swollen lump along their body, which is their recent meal bulging in the gut. When disturbed, many snakes will regurgitate their food, either as a defensive strategy or to help them escape. At the least, this can be a bother to the snake, and often is a loss of valuable energy. In general, when thinking about handling a snake or any other animal, remember to do so only if you can ensure your safety and the animal's safety.

Lizards also tend to bite when handled, which can be startling but rarely painful. Do not pick up a lizard by its tail. Many lizards and salamanders have adopted a peculiar defensive strategy in which they shed their tails when they are grabbed. Although this strategy may allow the animal to escape its predator, it can come at great expense. The energy they will need to invest into growing a new tail can take away considerably from their ability to reproduce, grow, or even survive.

A notoriously aggressive character is the snapping turtle. It is extremely ornery and should not be handled (or, at the very least, it should be handled with extreme care by someone with experience). If you must pick up a snapper, use a shovel to support the body and to keep you out of the reach of its bite. Also, do not lift it far off the ground, because it can move off the shovel easily. These are not the only turtles to beware of. In fact, all turtles can inflict a very painful bite, and many have sharp claws that will scratch you. For safety reasons, hold turtles securely by the upper shell, with one or two hands grasping the sides of the shell, while keeping fingers and hands well away from the head. Also keep the turtle's head aimed away from you so it can't latch on to a nearby part of your body. Never place your hands in front of the turtle's face or even close to its head. And remember, some turtles have very long necks and can reach around to bite.

The most useful safety tool is knowledge and familiarity with the animals in your local area. In some regions reptiles such as snakes, lizards, and alligators can be downright

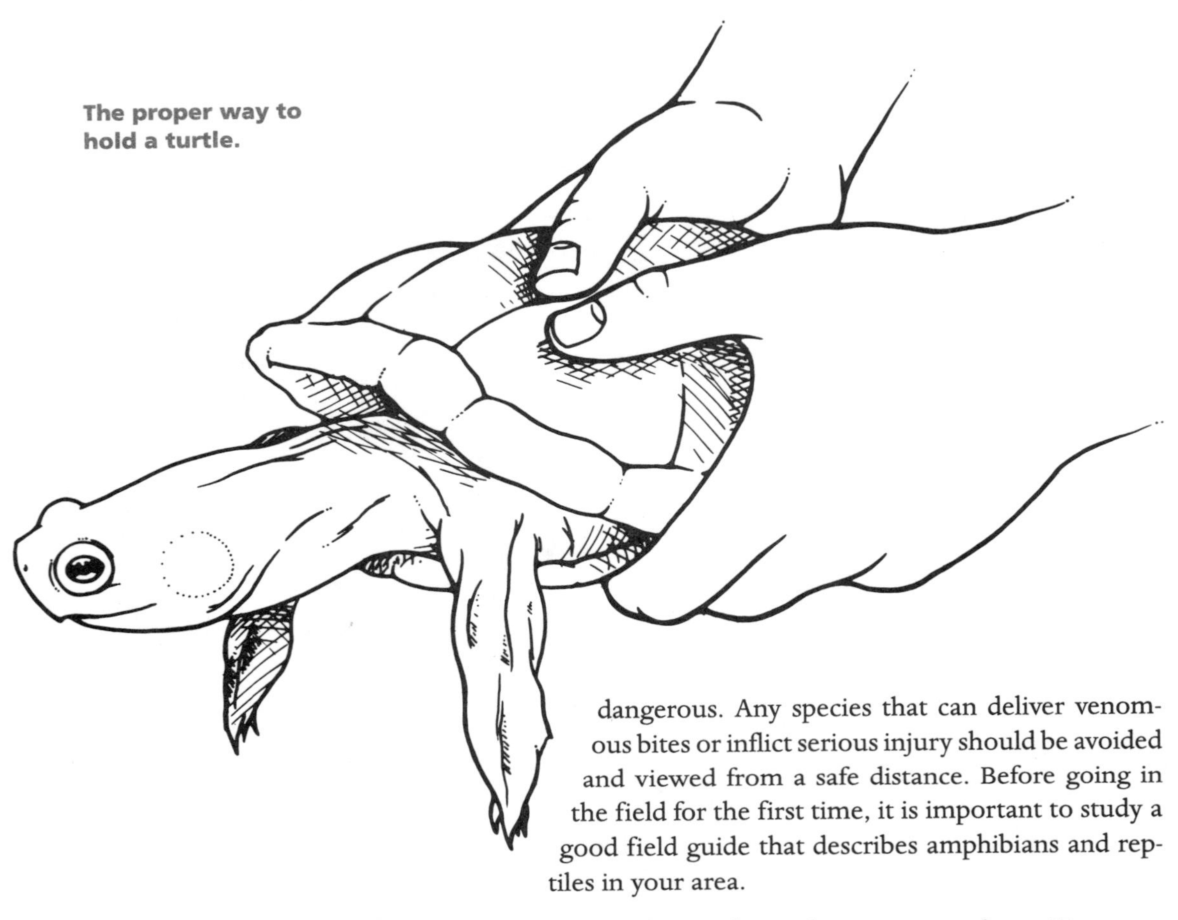

The proper way to hold a turtle.

dangerous. Any species that can deliver venomous bites or inflict serious injury should be avoided and viewed from a safe distance. Before going in the field for the first time, it is important to study a good field guide that describes amphibians and reptiles in your area.

Keep all herps out of the direct sun so they will not dry out or overheat. Herps are ectotherms, or animals that derive body heat from external sources. Usually they avoid overheating in their natural environments by immersing themselves in water, burrowing in the soil, or seeking refuge under leaf litter. Therefore, when restraining amphibians and reptiles you may want to provide objects for shade and moisture, such as damp leaves or soil. Also you should try not to hold them in your hands for too long. A small animal on a cold day can raise its temperature way above normal, simply by absorbing heat from your body. The best way to avoid causing stress from heat or water loss is to restrain the animals only minimally or not at all, and try not to alter their conditions too much from those of their natural environment.

After handling herps, return them to their natural habitats as soon as possible. Excessive handling and confinement stress these animals in many ways. Also, because herps have homes and preferred territories, it is important to return each animal to the place where it was found. Many amphibians and reptiles when displaced will undergo long and difficult journeys attempting to return to their original homes. Even if the release site seems acceptable, a relocated animal does not easily find a suitable home, and often is vulnerable to predators while searching for one. When returning an animal, try to return it to the same exact spot where it was found. If it was found under a rock or a log, don't place the object on top of the animal. Instead, replace the cover object to its original position first, then place the animal alongside, so it can easily and safely crawl underneath.

Safety Concerns

The good news about working with herps in the northern United States and eastern Canada is that only a few species are poisonous, and even those are relatively scarce. Thus the likelihood of interaction with a dangerous herp is low, and in many areas negligible. Nonetheless, as is true when handling any live animal, you should use caution and common sense. In the southern states, there is a much greater variety of dangerous reptiles and you should be especially cautious and knowledgeable before attempting to handle them.

The basic rules of herp safety are (1) do not pick up any animal unless you need to, and (2) know the animal before you pick it up. Before making a move, you should be able to identify the species and be somewhat familiar with its biology and behavior. Most important is the ability to identify poisonous snakes in both their juvenile and adult stages. Do not pick up any snake if you are uncertain whether it is poisonous! And remember that even turtles and nonpoisonous snakes may deliver a painful bite if frightened or mishandled.

Out in the field, use the following commonsense guidelines to avoid an unwanted interaction with a snake or other herp. When looking under objects, always lift them with the exposed side pointing away from you. That way, if a snake is present, it will not be facing you directly. Always check the surface of rocks or logs before leaning or sitting down on them. If you are walking through terrain likely to have venomous snakes, walk slowly and look at the ground while you are walking. Scan ahead and around you for any snakes within a radius of approximately 2-3 m (7-10 ft). Don't worry about snakes at greater distances because this distance gives you plenty of time to react. Stop and look around for snakes and listen before becoming engaged in a conversation or in observing plants, rocks, or other animals.

Although avoiding poisonous snakes is the best defense, be prepared in case of a snakebite. First, in advance of your field trip, locate a hospital equipped to deal with snakebites. Discuss with a trained medical technician the best method of action to take if someone in your group is bitten. Also, come up with a plan beforehand for transporting someone to a nearby treatment center. People who are well informed and have a prearranged plan always make the best responses to snakebites.

Next, expect the victim of a snakebite to experience pain. Being bitten by a snake can hurt a lot, even when no venom is injected—imagine the pain of a sharp nail driven into your hand, arm, or leg. In any circumstance, do not panic. If the snake was poisonous, the victim should have half an hour or longer before there are serious effects. Keep the victim quiet and calm, if possible. Try to have the victim exert minimal effort while you are getting him or her to a vehicle and then to a nearby treatment facility. It is very important not to overreact or to waste time in often fruitless, and sometimes dangerous, self-administered treatment. Get to a treatment center and let a professional take it from there. If there is no suitable treatment center in your area, it is essential to learn much more about venomous snakes and forms of treatment in advance.

Bites from snakes and turtles are not the only potential sources of concern. Some toads, frogs, and salamanders exude poisonous fluids from glands in their skin. These secretions generally do not affect human skin, but the moist membranes of your eyes, mouth, and nose are extremely sensitive. Therefore, do not wipe your eyes, nose, or mouth after handling toads and other herps. In addition, always wash your hands before eating or drinking. Antibacterial soap or wipes can come in handy for this purpose.

Be prepared for other dangerous and annoying creatures in the field. Bee stings and spider bites are a more likely threat than snakebites. If someone in your group is allergic to bites and stings, be sure to carry appropriate medication. You can use bug spray where mosquitoes, flies, or ticks are common. You don't need to place it directly on your skin but instead can spray it on a hat or neck scarf and wear a long-sleeved shirt. Lyme disease, a serious bacterial infection transmitted by deer ticks, is a growing problem throughout some parts of the United States. If you are working in an area known to have deer ticks, keep as much skin covered as possible. Wear long-sleeved shirts and tuck your pant legs into long socks. Upon returning from the field, check yourself thoroughly for the tiny ticks before they have a chance to bite. Remove your field clothes before entering a living area and place them into a bag or directly into the laundry. Drying clothes at a high heat setting for 20 minutes will kill the ticks.

Also be aware of toxic plants when hiking in the outdoors. Poison ivy is well known, but stinging nettles and other plants may be present as well. You may want to invest in a good plant guide and learn the characteristics of the potential problem plants in your area.

Prior to working with herps, you should become familiar with your local school system's safety regulations, standards, and policies. Boards of education often have written policies relative to use of animals in the classroom and out in the field. Individual school practices may also address your responsibilities for the safety of students during laboratory or field experiences. The school nurse should be contacted prior to field experiences to determine requirements for being aware of students with allergies or other medical concerns and for administering medications. If you are a middle or high school science teacher, you should review your science department's Laboratory Safety Standard Plan (Chemical Hygiene Plan required by OSHA in most states) as an additional safety resource.

When preparing for the field, take sensible precautions, including wearing a hat and using sunscreen to avoid sunburn, carrying a water bottle, and wearing socks and solid shoes that cover your feet. Also let someone know where you are going ahead of time, and when you plan to return. Many field activities take place along the edges of ponds and streams; thus, good water safety practices should be followed. Expect to get wet and dress appropriately. Young children should always be supervised by an adult, and in some areas, it may be appropriate for youths to wear life jackets.

By taking a few precautions beforehand and exercising care during your excursion, you and your youth group can have great fun in the field—and learn a lot, too!

Permits and Regulations

Many states require permits for collecting vertebrates. Contact your state natural resources or environmental conservation agency to determine what regulations apply to the activities you are planning. If a permit is needed, the application process may take several weeks to months, depending on the status of the species in which you are interested.

Some animals are highly protected, either at the state or federal level. At the least, these animals usually require special permits that only allow very specific activities. If an animal is listed as threatened or endangered, generally no disturbance of any kind is permitted. If there is any question whether an animal is a protected species, it is best to leave it alone in its natural state.

If you are buying species from a pet store or dealer, you should ask the owner or dealer whether the species is protected. Some of the animals may have been obtained illegally, so it is important to inquire how the animals were obtained. This may be the case for native as well as exotic species. Finally, do not return purchased animals to the wild when you are finished studying them. Instead, they should be kept as pets or donated to a nature center or school. Serious problems can occur when animals are artificially introduced into natural populations where they can become sources of disturbance and disease.

Basic Facts about Herps

The study of amphibians and reptiles is called herpetology. The term is derived from the Greek word *herpeton*, meaning "creeping thing." Herpetology is unlike the study of mammals (mammalogy) or birds (ornithology) because it considers together two very different classes of organisms, instead of just one. The historical lumping together of these very different groups was probably influenced by the perception of them all as ground-dwelling, secretive, egg-laying, and "cold-blooded" creatures. Our current understanding, however, is that reptiles and amphibians differ radically in their structure, development, and natural history. In fact, scientists now believe that reptiles have much more in common with birds than they do with amphibians! Therefore, in many ways, the grouping of amphibians and reptiles as "herps" is somewhat artificial and outdated.

It wasn't until well after the development of a scientific system of classification that amphibians and reptiles became recognized as vastly distinct animals. Our current classification or "taxonomic" system started with Carl Linnaeus in 1735. His taxonomic scheme provided a powerful tool to identify and name species based on their biological characteristics and their relationships to each other. After more than two centuries of study and refinements to Linnaeus's system, our current classification system places amphibians and reptiles into distinct classes, reflecting their extreme diversity of lifestyles and biological traits.

For most of the 20th century, our thinking was that at higher levels of classification, all herps fall within the kingdom Animalia, the phylum Chordata, and the subphylum

Vertebrata, that is, those animals that have some sort of backbone during their development. Among the vertebrates, fish, amphibians, reptiles, birds, and mammals were considered to be major classes. Although the current classification system is pretty similar at the higher levels, scientists have made many recent changes as a result of our new understanding of relationships among vertebrates. The familiar traditional classes of fish have been further divided, and birds are now thought to be a subgroup of reptiles, often referred to as "reptiles with feathers." One thing that has remained steadfast, however, is the clear distinction between amphibians and reptiles.

Differences between Amphibians and Reptiles

At the heart of the distinction between the two herp groups is the nature of their eggs. Amphibian eggs are much more like those of the primitive vertebrates. In fact, the gelatinous eggs of amphibians represent very little change from those of their fish ancestors. A single membrane surrounds the yolk and developing embryo throughout incubation. This central egg is surrounded by a series of jelly-like capsules that help protect the embryo and buffer it from the outside world. Although the gelatinous capsules contribute greatly to keeping the egg from drying out, many amphibians can only breed in water, and those that reproduce on land must still find moist places to nest.

As a result of evolutionary changes in amphibian eggs, the reptiles became the first vertebrates to live fully on land. The reptile egg was revolutionary, in that it was the first to have a series of four membranes that compartmentalize the egg, allowing transport of nutrients, essential elements, and wastes to and from the embryo. Importantly, the reptile eggs are covered in an outer membrane, and many species further protect the eggs in a leathery shell. These outer coverings keep the eggs watertight, which has allowed reptiles to expand into all sorts of environments, including the driest deserts. If the features of a reptile egg seem familiar, that's because they are shared almost to the letter by birds and, believe it or not, mammals.

Amphibians and reptiles differ in many other ways. *Amphibian*, which comes from the Greek word *amphibios*, meaning "leading a double life," is an appropriate name for this group of animals that have moved onto land but remain very much connected to water. Evolutionary biologists describe the amphibians as transitional between fish and terrestrial reptiles. Most believe that a group of prehistoric fishes first developed legs, then with a few modifications to their lungs and hearts, became the first amphibians. They never completely broke their dependency on water, however. This shows up in two important ways. The first is in reproduction, since their eggs must remain wet or moist throughout development. The second is in their skin, which is smooth and relatively thin. Most amphibians supplement their breathing by transporting oxygen through their thin skin and into blood vessels lying just underneath. For this to be most effective, they must remain in moist or wet environments.

Reptiles, in contrast, are very well designed for life on land. Along with the self-contained and weather-resistant egg, their bodies are covered with a layer of armor in the form of scales. The scales are made of the same materials as the feathers of birds and the toenails of mammals. This outer covering protects them and prevents water loss, which allows them to live their entire lives out of water. After reptiles first evolved from a specialized group of amphibians, they very rapidly expanded into all kinds of species and moved into all types of different habitats. For many millions of years they remained the only vertebrates in many places on land. Indeed, it wasn't until birds and mammals came along that reptiles had any real company in dry places.

Along the way, many reptiles made some pretty important improvements to their legs, which allowed them to run and better support their weight. They also were able to breathe more efficiently by expanding their chests and lungs. (Their amphibian ancestors mainly pump air into their lungs by squeezing mouthfuls of air down their throats.) Because of these and other important modifications, reptiles were able to reach larger sizes than amphibians. The biggest and most infamous group of reptiles was the dinosaurs, but even today, some lizards, turtles, snakes, and crocodile relatives are impressively large.

Scientists have pieced together the intricate relationships among amphibians and reptiles by examining fossils and by comparing living species. Fossils found in rocks over 370 million years old have shown us the fishlike appearance of one of the very early amphibians, supporting the idea that these creatures evolved from prehistoric fish. As we examine the complete fossil record, we see that over the next 100 million years or so, amphibians blossomed into a group of more than 40 families, becoming very abundant in swamps and on land. Sometime after that, however, the amphibians became greatly reduced in numbers. Currently we have identified about 4,000 species worldwide, and these belong to three major groups.

The first group, frogs, accounts for the vast majority, with over 3,500 species. There are about 100 species of frogs and toads in North America. Frogs are four-legged amphibians, well designed for hopping or jumping. They all have lungs for breathing, and many have vocal chords they use to call out loudly to attract mates. The words *frog* and *toad* are not technical terms, but are commonly applied to animals that have slightly different characteristics. In general, toads move around using small hops, while frogs tend to jump and leap.

The modern salamander group is much smaller than the frog group, with only about 390 species. The center of evolution of many salamanders appears to have been North America, which contains more than one-third of all the world's species. Salamanders are smooth-skinned, elongated amphibians that usually are very quiet and secretive. Most species are four-legged, but a few have greatly reduced limbs or no hind legs. One of the most numerous groups is the family of lungless salamanders, many species of which breathe entirely through their skin. Salamanders are widespread throughout North America, occurring in lakes, streams, swamps, and forests.

The third group of modern amphibians is a weird and mysterious group of legless, wormlike, nearly blind animals called caecilians. There are around 165 species worldwide, distributed throughout most tropical areas, including Central America. These burrowing animals look like worms, but upon close inspection it becomes apparent that they have eyes and a mouth with teeth. Since they live underground, we know little about the activities of caecilians.

The reduction in diversity of amphibians coincided with a great increase in reptiles beginning around 250 million years ago. According to the fossil record, after 50 million years of evolution, a group of amphibians underwent some changes that led to the first reptiles. It is from these primitive reptiles that many different forms evolved, including mammals, modern-day reptiles (such as lizards, crocodiles, and turtles), and a group that later branched off into dinosaurs and birds. These changes were by no means quick. During that first 50 million years or so, while amphibians were much more numerous, early reptiles slowly developed different body shapes and strategies for walking, swimming, and even flying. As their variety increased, their numbers also appeared to grow, until they became the most abundant vertebrate group on land. This was the beginning of the age of the dinosaurs, one of the most exciting times in vertebrate history. Dinosaurs were a successful and diverse group that dominated land for about 150 million years before they all became extinct rather abruptly.

These may seem like unimaginably long periods of time. But think of the crocodilians (e.g., crocodiles and alligators) and turtles, which have persisted through millions of years as the early reptiles came and went, and as the dinosaurs blossomed and then winked out of existence. They have continued to plod along, seemingly unaffected, as some of their relatives grew fur or feathers and went on to dominate the land and the skies. To us, these represent awesome events that occurred over immense spans of time. But when compared to turtles and crocodilians, which have been around a couple hundred million years, these developments appear to be no more than a series of fleeting events.

Today, the reptiles still reflect their varied and long ancestry, with over 7,000 species representing major groups that differ in some pretty radical ways. Joining the armored forms of the turtles and crocodilians are the lizards, snakes, and the lesser-known tuataras. The tuataras are the last remaining two species of a once diverse group of lizardlike reptiles. They only occur on several islands off the coast of New Zealand.

In contrast, there are more than 3,500 species of lizards. These include some well-known types, such as iguanas, geckos, and chameleons, along with some obscure forms, such as the legless worm lizards. Lizards are worldwide in distribution, extending from the tropics to very high latitudes. The largest group among the lizards is the skink family with over 1,000 species. Lizards have come up with a variety of strategies for movement and other life processes. Many race around in high-speed bursts; some even run on two legs. A specialized tropical lizard can run upright across the surface of the water. Some lizards are specialized burrowers that spend much of their life underground, while others have special flaps that allow them to soar through the air over great distances. In addition, many have adopted different techniques to reproduce, such as laying eggs, bearing live young, and even a form of asexual reproduction whereby females virtually produce clones of themselves.

Snakes also are widespread and highly diverse, with around 3,000 species worldwide. They are all without limbs, a feature that also is shared by some of their lizard relatives. Snakes are unique in their ability to move their upper and lower jaws independently. This allows them to open their mouths very wide to swallow large prey, often wider than their own bodies. Some snakes swallow their prey whole while others, such as boas, first squeeze their prey by tightly coiling around them. An entire group of snakes injects venom into their prey, by squeezing poison out of special sacs and delivering it through punctures caused by their fangs. Many snakes reproduce by laying eggs, but several groups bear live young, including the rattlesnakes. Snakes vary widely in size, ranging from species that are not much larger than an earthworm, to the giant anacondas that can reach lengths as great as 10 m (33 ft).

The fame of the crocodilians seems to be a bit out of proportion to their actual presence in today's world. Perhaps the image of their powerful jaws, or their quick and violent bursts of speed, captures our imagination. Or perhaps it is the vision of a crocodile silently slipping off the riverbank into the water in deadly pursuit of unwary prey. Nevertheless, today there are only 22 species of crocodilians, all of which largely are confined to the tropics and subtropical regions. Crocodiles, alligators, and gharials (which are represented by only one species) are all that remain of the prehistoric crocodilians.

They all are powerful swimmers, with a muscular tail that propels them through the water. They lay eggs in mounds that the female constructs out of vegetation and soil. All crocodilians show a high degree of parental care. Mothers guard the nests and help the hatching process. In some species, particularly among the alligators, the mother continues to protect the young for a year or more, and sometimes they live in extended family groups for even longer.

Among all of the vertebrates, the turtles are an obvious standout. Their special features, such as an upper and lower shell made of bones, are truly astonishing developments among the vertebrates. The success of turtles is demonstrated by the fact that after more than 200 million years of existence, they remain relatively unchanged. Some turtles can withdraw their head and legs and completely enclose them within their shells, sealing themselves off from the outside world. Many are docile, subdued creatures, while others are aggressive and easily provoked. Today there are more than 250 species spread throughout all the continents, ranging from the tropics to high latitudes. Turtles occupy many habitats, from the tortoises in the driest desert areas, to the semiaquatic turtles inhabiting freshwater lakes and swamps, to the sea turtles that travel the world's oceans. All turtles are egg-laying reptiles, depositing their eggs in nests on land that they scoop out with their hind feet.

Similarities between Amphibians and Reptiles

Despite all of the differences between and among amphibians and reptiles, a single feature that traditionally united them in our minds was the notion of their being "cold-blooded." Today, we are not as keen on this term as a description of these animals. In fact, many reptiles can maintain body temperatures as high or higher than humans. Think of a small lizard resting on the sand of a desert or on a hot beach. The body temperatures of these animals are far from cold.

Two sets of terms help us better describe the temperature regulation of amphibians and reptiles. They have to do with how animals warm or cool themselves and how variable their body temperatures are. Animals whose temperatures are regulated by outside sources in the environment are said to be *ectothermic*. This is different from *endothermic* animals, such as mammals and birds, whose temperature is primarily generated from

SCI LINKS
THE WORLD'S A CLICK AWAY
Topic: **temperature regulation**
Go to: **www.scilinks.org**
Code: **HERP11**

internal body processes. Mammals and birds also closely regulate their range of internal body temperature, not letting it vary by more than a few degrees during normal activities. This tight control is called *homeothermy*. Most amphibians and reptiles, however, are able to withstand wide ranges of body temperatures and still continue to function. This is called *poikilothermy*.

The ability to withstand such a broad range of internal body temperatures is a common trait shared by many amphibians and reptiles. Sometimes this thermal flexibility can be truly impressive. The same small lizard that exposes itself to hot desert sands also can experience very cold nights and still remain active. The same painted turtle that basks in the warm summer sun also can be seen swimming slowly beneath the ice in the middle of winter. Similarly, it is common to see hundreds of newts swimming around in frozen lakes of the far north.

These and other shared features connect amphibians and reptiles and go some way toward justifying the traditional lumping of these distinct groups of animals. But far beyond shared ancestry and some similarities in behavior and physical structures, the amphibians and reptiles will always be linked in their ability to excite and fascinate children and adults alike.

Biology: Up Close and In Hand

Herps have adapted to their environment through evolving many unique biological structures and strategies. This section considers several fascinating aspects of their biology, including their form, skin, defensive systems, reproduction, and development. There are 11 activities that provide opportunities for hands-on experience and learning. Activities 2.1, 2.2, 3.1, 3.2, 4.1, 5.1, and 6.3 can be simplified for use with younger children. The remaining activities are more appropriate for youth age 12 and up.

Chapter 1. Morphology—Getting into Shape(s)

Chapter 2. Skin—It's Packaging Plus

Chapter 3. Defense Systems—On Guard!

Chapter 4. Multiplying Herps—Reproduction and Development

Chapter 5. Breeding Behavior

Chapter 6. Life Span and Life History

Chapter 7. Herps through the Ages—Evolution and Extinction

Chapter 1

Morphology—Getting into Shape(s)

National Science Education Standards
Grades 5–8: Structure and function of living systems (Life Science)

An animal's body must simultaneously serve a variety of functions, including defense, locomotion, feeding, and reproduction. By examining the body structure of different herps, we can see the relationship between the shape or form of a part of an animal's body, and its use or function.

A classic example of the relationship between form and function is the tanklike shell of a box turtle. When threatened, the turtle withdraws its legs and head, enclosing them within its shell. Foxes, raccoons, and other predators are quickly discouraged by this bony barricade and soon turn away. The box turtle is left to continue on its slow but safe travel.

Protection is not the only consideration. Every herp must have a means for moving from one place to another. It is amazing to observe the diversity of herp body forms that have evolved to meet this common need for movement. As a group, herps use almost every form of locomotion, including crawling, swimming, climbing, hopping, and even gliding through the air. For some herps, such as sea turtles, the limbs make up a large portion of the body and are the primary means of locomotion. In others, the limbs are hardly used or even may be absent, as in many lizards and snakes. Despite this variation, all reptiles and amphibians have evolved from a basic tetrapod (four-legged) structure. Even in the herps without limbs, evidence of their common ancestry often can be seen when you carefully examine their skeletons.

The large hind legs of frogs are ideally suited for hopping and swimming. Bullfrogs have powerful leg muscles and have been known to jump more than 8 m (26.4 ft) in a single hop. Webbing between the toes of their hind legs also helps push the frogs through the water. Although the seaturtle looks very different from the bullfrog, it too has evolved a body form suited for swimming. The seaturtle's flippers and streamlined body allow it to swim effortlessly between 20 and 60 km (12–36 mi) per day.

A treefrog's limbs have adaptations for climbing. These frogs have suction cups on the ends of their toes that help them attach to and climb up vertical surfaces. It is not unusual to see a gray treefrog clinging to the outside of a window at night, hunting insects attracted by the light.

In contrast to most other herps, snakes do not have external limbs, although most primitive snakes do have traces of hind limbs or a pelvis as part of their skeletons. A legless body helps snakes maneuver into small holes and through dense vegetation. Many snakes move quite fast by repeatedly sending a muscular wave down the length of their bodies. Ratsnakes even use their wavelike motion to climb tree trunks to reach bird eggs. A similar rippling motion allows many snakes to swim rapidly across a pond or marsh. This way of life is so successful that other herps, such as glass lizards and sirens, have developed legless body forms similar to that of the snake.

Herps also have evolved a vast assortment of specialized adaptations that allow for different feeding strategies. Alligators have large pointed teeth and strong jaws to secure their food. Turtles have similarly powerful jaws but they have a sharp hardened beak in place of teeth. Some salamanders have long, sticky tongues that dart out with lightning speed to distances as long as half their body length. Frogs have tongues that are attached in the front and actually catapult out of the mouth to snare unsuspecting insects. Perhaps the most notorious examples of special feeding adaptations are the venom-injecting fangs of some snakes.

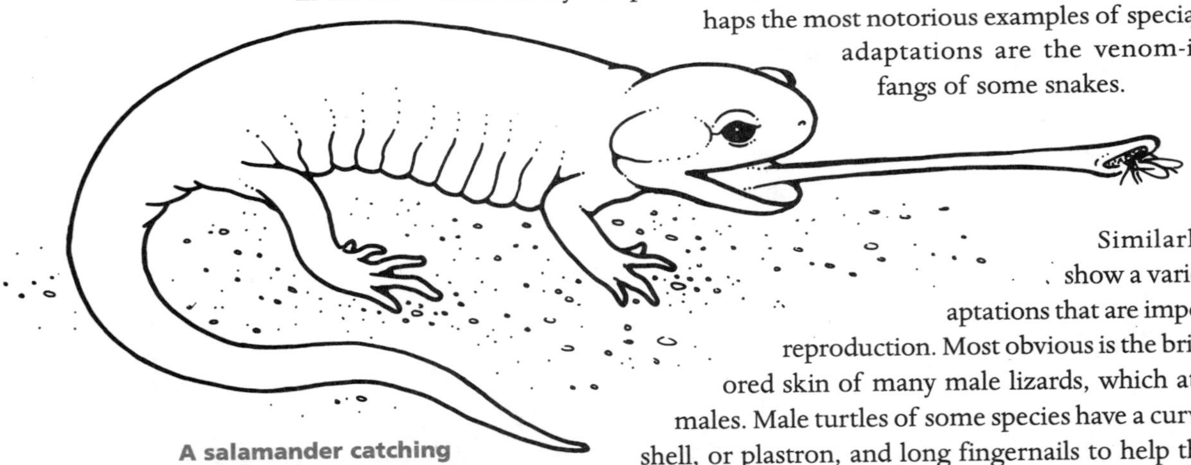

A salamander catching its prey with its long, sticky tongue.

Similarly, herps show a variety of adaptations that are important for reproduction. Most obvious is the brightly colored skin of many male lizards, which attracts females. Male turtles of some species have a curved lower shell, or plastron, and long fingernails to help them hold on to a female during mating. Some male frogs and toads have hardened pads on their thumbs for similar purposes. Differences in size between the sexes also are common in herps. In many species, the female is larger than the male in order to allow room for the eggs developing in her body. In other species, the males grow larger to successfully fend off other males competing for mates.

Observing the adaptations of a herp provides useful insights into its way of life. You have just seen how shapes, structures, and sizes of herps vary. In upcoming chapters, we will examine many other physical and behavioral adaptations—all of which are important to the survival of these animals as well as making them interesting objects to observe.

Activity 1.1. Herps Inside-Out

Objectives

Students will understand (a) the relationships between body form and function in several herps and (b) how different groups of herps have modified their skeletons to allow different movements.

Materials

- skeletons and/or parts of skeletons of snake, turtle, and frog (Clean, preserved skeletons can be obtained from several biological supply companies. Also, recent road-kills can provide a unique opportunity to examine local species; these specimens, however, will need some cleaning and freezer storage.)
- hand lenses
- copies of labeled skeletons (provided)

Procedure

1. Distribute the photocopies of the labeled skeletons and real skeletons if available.
2. Have students use the labeled drawings to identify different bone structures. Ask them to compare the limb and body structures among the different animals.
3. Discuss how the bones and other structures (e.g., turtle shells) have been modified relative to their functions for locomotion and defense. Note the following comparisons:

- The sternum, or breastbone, is totally missing in the snakes but has spread out to form the entire lower shell, or plastron, in the turtle.
- The upper and lower jaws of the snake are separated to allow them to swallow large prey.
- The snake has multiple vertebrae (85–400) in contrast to the turtle, whose vertebrae are fused to form the upper shell, or carapace.
- The hind leg bones and toes of the frog have become extremely elongated to allow for hopping and swimming.

4. Challenge your students to come up with other observations and comparisons related to defense, locomotion, feeding, and reproduction.

Skeleton of a snake

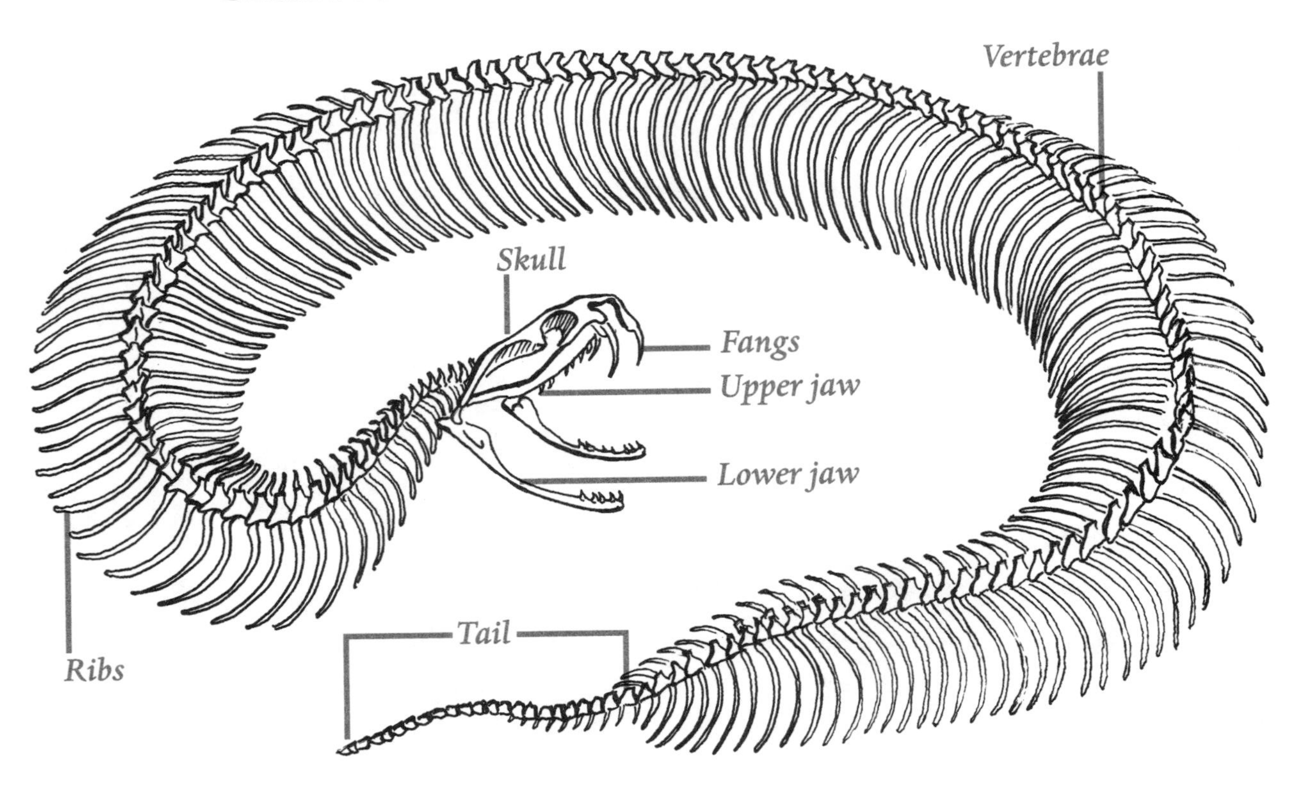

Skeleton of a frog

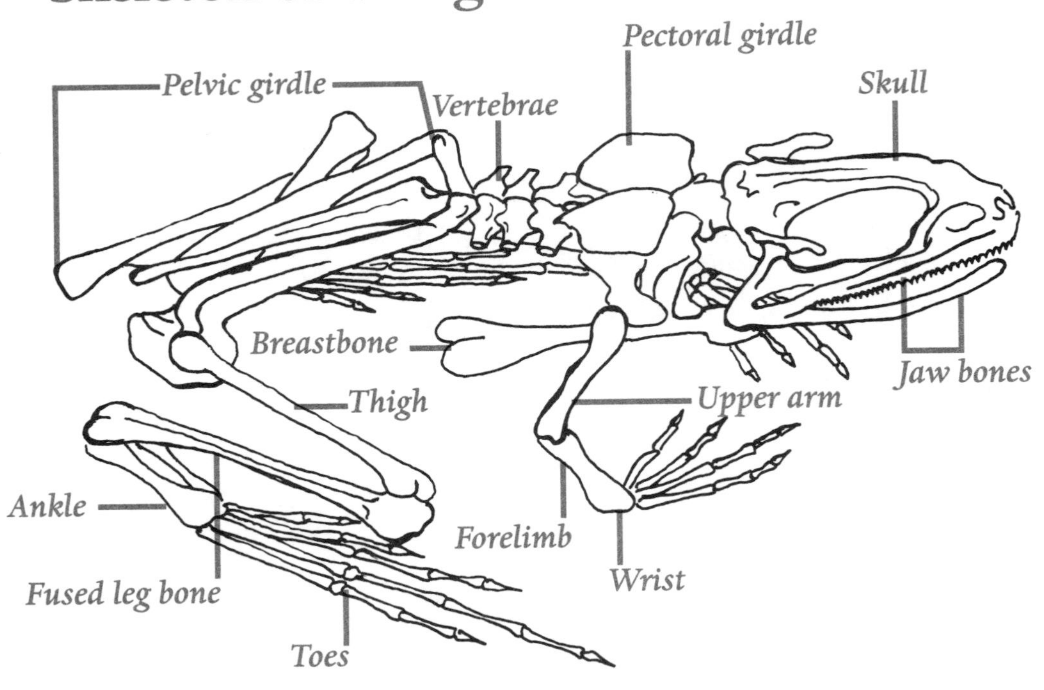

Skeleton of a turtle

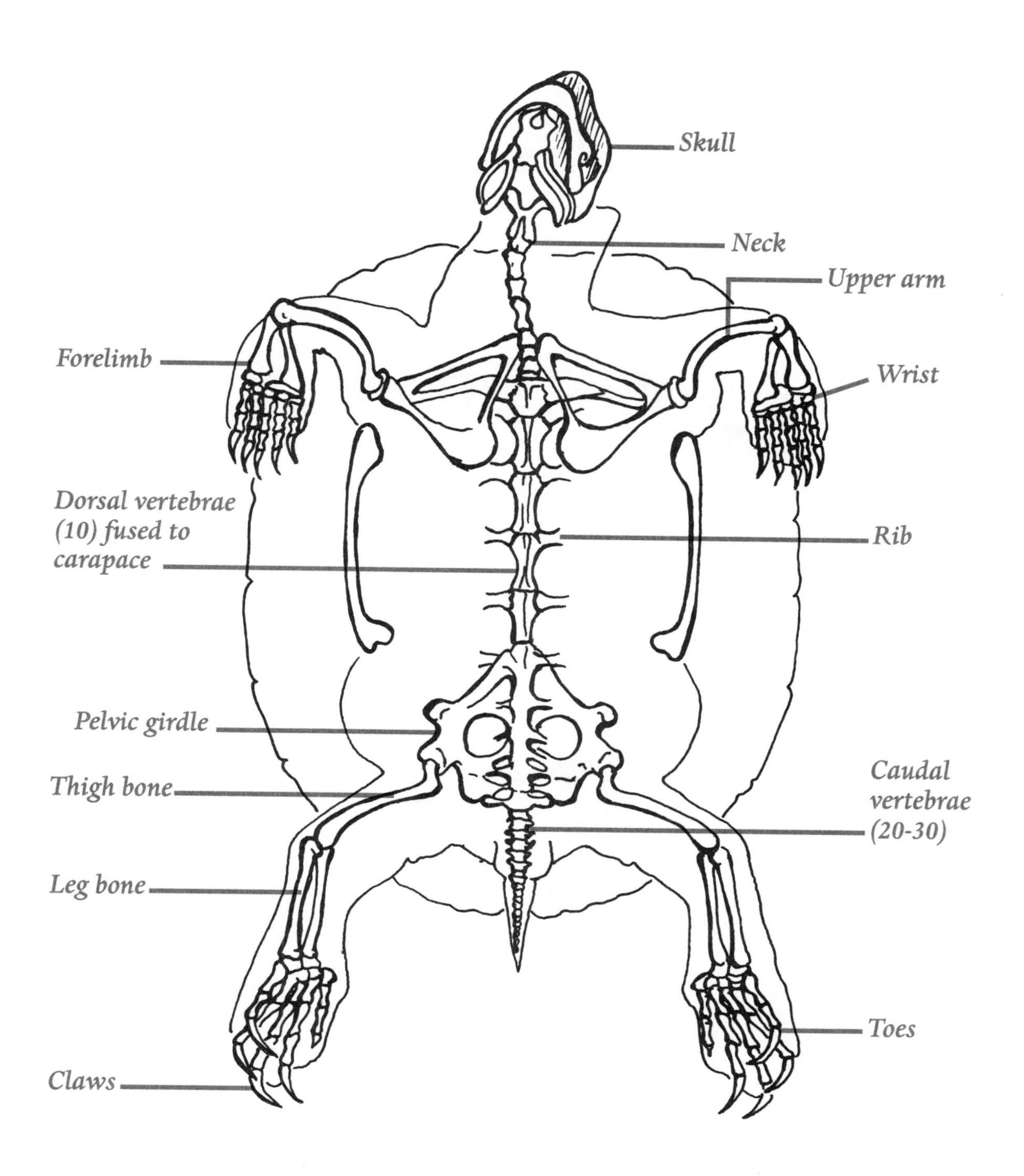

Chapter 2

Skin—
It's Packaging Plus

National Science Education Standards
Grades 5–8: Structure and function of living systems (Life Science)

SCi LINKS
THE WORLD'S A CLICK AWAY
Topic: **herp skin**
Go to: **www.scilinks.org**
Code: **HERP21**

Although it is easy to take the outside of an animal's body for granted, it performs a surprising variety of functions. In fact, skin is one of the most remarkable parts of a herp's body. You can easily observe fascinating differences in skin color, texture, and function when comparing amphibians and reptiles.

The skin of a snake is characteristic of many reptiles. It consists of hard scales connected by thinner, more elastic skin. The scales are shiny and smooth, and not at all slimy. Thousands of scales overlap and provide a flexible system that bends easily with the snake's rippling movements.

The scales on the underside, which are wider than those on the snake's sides and back, are tough enough to withstand constant rubbing as the snake moves across rough ground and branches. In addition, the elastic skin between the scales can stretch up to four times its normal size when the snake is digesting a large animal that it has swallowed whole.

Throughout its life, the snake periodically sheds the outer layer of its skin to allow room to grow. The skin is discarded as a single piece, from the tip of the nose to the end of the tail. The discarded skin also includes the eye covering, or ocular shield, which snakes have developed as a modification of eyelids. Interestingly, rattlesnake rattles are modified scales that grow at the end of the tail. These do not come off when a rattlesnake sheds its skin. Instead, after each shedding, a new segment grows underneath the previous one, creating a chain of rattles locked together like a set of child's pop beads.

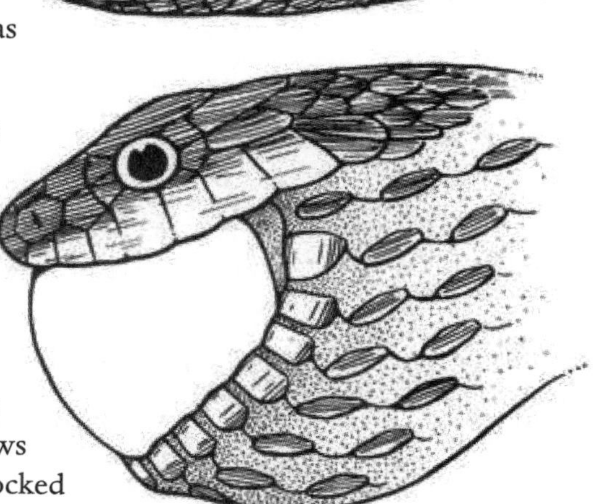

Head of a snake with scales as usually seen and with scales stretched by elastic skin to allow it to swallow an egg.

The skin of each herp species has a unique color and pattern, which come from special pigment cells beneath the skin's surface. Color can provide camouflage, or even warn predators and other animals. Gray, green, and brown markings of a treefrog form a mosaic that helps to camouflage it against a leafy background. Coralsnakes and newts have bright, contrasting markings of orange or yellow that are most likely used to warn potential predators of their toxicity. The colorful stripes and spots of some turtles can help them distinguish each other, especially during mating. Color differences also can help animals distinguish males from females and adults from juveniles of the same species.

Skin color also can be important in controlling body temperature, or thermoregulation. Most herps are ectotherms, that is, the temperature of their surroundings influences their body temperature. Therefore, they generally rely on heat from the environment for survival. On a cold day, a dark-colored herp can absorb the sun's heat very quickly. Many amphibians are capable of changing color and even getting darker to absorb more heat and stay warm as it gets colder. This quest for heat explains why snakes often are seen basking in the open and on warm roads.

Most amphibians also use their skin as a means of absorbing oxygen. In fact, many species do not have lungs, including the brook, red-backed, and dusky salamanders, and respire only through the skin's surface. Just below the thin outer layer of skin are numerous blood vessels that aid in respiration. For respiration to occur, the skin must remain moist at all times. This is one reason most amphibians flourish in moist habitats such as streamsides, wetlands, and the upper organic layer of forest soil.

Reptile skin, on the other hand, is thick and dry. Its most important function is the conservation of water within the animal. The thick layer of compact and hardened scales also provides an excellent protective armor against toxic chemicals, physical damage, and predators.

Activity 2.1. Skin Textures

Objectives

Students will (a) learn the proper methods to handle live herps, and (b) understand variations in the skin of different groups of herps.

Materials

- live specimens of at least one amphibian, such as a frog, toad, or salamander. You may be able to arrange with a nature center to hold and observe these animals there.
- snake skin or live specimen of a nonpoisonous snake.
- copies of "Care and Handling of Live Herps" (pages 2–5)

Procedure

Use the instructions in "Care and Handling of Live Herps" to explain the proper way to hold different kinds of herps.

Safety Tip: Hold live herps gently but firmly, supporting their bodies and immobilizing their heads, to reduce chance of biting.

1. Let students handle the live organisms and snake skin. Have them record their observations of the differences between the skin of the different organisms. Is the skin slimy or scaly? How do the scales on the different parts of the snake's body differ?

2. Ask the students how the different types of skin might relate to their functions or particular roles. For example, what role does the skin play in locomotion, defense, metabolism, and breathing? Discuss the relationship between the habitat requirements of amphibians and reptiles and the different types of skin they have.

Activity 2.2. Lurking Lizards

Green anole lizards (*Anolis carolinensis*) are common in low shrubs and other plants of the Southeast and as pets elsewhere in the United States. They sometimes are mistakenly called "chameleons" because they are able to change color from green to brown to match the surrounding vegetation. However, true chameleons come from Africa, Madagascar, and southern India.

Objective

Students will understand one means of camouflage used by chameleons and anole lizards.

Materials

- pet anole lizard(s), available from pet store
- cage/aquarium with screen lid and some branches or small plants
- squares of colored cardboard at least 15 cm (10 in) on a side. (Select some colors that the animal might encounter in its native habitat—e.g., greens and browns—and others it would not normally encounter—e.g., orange and blue.)

Procedure

1. Set up the aquarium where the group can easily observe it and give the lizard some time to become comfortable in its new surroundings.

2. Have the students describe the coloring of the lizard and record their observations.

3. Place a square of colored cardboard in the cage at an angle so that the lizard will comfortably perch on it or nearby.

4. Gently place the lizard on the square. This may take some patience as the lizard may move off.

5. Have the students record the time and the color of the cardboard. They should observe the color of the lizard at regular intervals (e.g., once every two minutes) and record any changes in color. They should record the time that the lizard left the square and whether they observed anything that might have caused it to leave.

6. Repeat steps 3–5 with other colors.

7. Have the group discuss their results. They can summarize what happened with each color and compare the differences in the animal's response to the different colors. How much did the lizard's color change? Was it able to match some colors better than others? Why might this be so? How quickly did the skin color change? Did the lizard leave some colors more quickly than others? If you had more than one lizard, were their responses different?

8. Ask the students to speculate on reasons for the differences they observed. How might they test these ideas further?

Chapter 3

Defense Systems—On Guard!

National Science Education Standards
Grades 5–8: Structure and function in living systems (Life Science)
Regulation and behavior (Life Science)
Grades 9–12: The behavior of organisms (Life Science)

Frogs, salamanders, snakes, and other herps are often small and live on the ground or in the water. Because of these characteristics, they are vulnerable to being preyed on by all kinds of carnivorous animals. In order to avoid being eaten, herps use a variety of strategies and protective mechanisms.

SCI LINKS
THE WORLD'S A CLICK AWAY
Topic: **animal camouflage**
Go to: **www.scilinks.org**
Code: **HERP25**

As a first line of defense, most herps try to avoid being seen by their predators. Many are nocturnal and use the cover of darkness to avoid notice. During the day, most herps tend to remain hidden beneath dead leaves, rocks, and logs, or in underground burrows.

Herps also avoid confrontation through camouflage. Using a variety of grays, greens, and browns, these animals can blend remarkably well into the background of their natural environment. It is amazing how difficult it is to see a smooth greensnake that is moving through the grass!

Countershading is an interesting form of camouflage for herps that live in the water. Many turtles, frogs, and salamanders have light colors on their bellies and dark colors on their backs. This color pattern makes them less visible to

A spring peeper with the "X" on its back that visually disguises its frog shape.

aquatic predators that see them against a light sky. Birds and other predators hunting from above also have a hard time spotting them against the dark water. Even some of the larger predators, such as snapping turtles and alligators, have countershading, perhaps to be less visible when stalking their prey.

A lot of species use spots, stripes, and blotches to break up the outline of their bodies when viewed against leaves or soil. The distinctive "x" on the back of the spring peeper is an example of a mark that allows this frog to virtually disappear when on the ground or perched on a blade of grass. Unlike animals that use camouflage, the colors of these animals do not necessarily blend with the background. In fact, many times the markings are quite bright and even gaudy. The eyes of the predator, however, are tricked into thinking the shape they are seeing is not an animal.

Some herps do not avoid or hide from predators but instead frighten them off by displaying warning signs. For example, toads and newts have glands in their skin that produce toxins. In order for this toxicity to protect an animal from being eaten, the predators must be reminded that they are about to eat something that will make them sick. A common method of alerting a predator is by being very brightly colored. This explains why the young newts, or efts, that we see walking around the forest are bright orange and yellow. Their color is a vivid advertisement of their toxicity. Other common examples of this aposematic, or warning, coloration are the brightly banded, venomous coralsnakes and the very decorative, poison dart frogs of Central America.

Interestingly, a herp truly may be poisonous or it may be just bluffing. Some harmless herps have adapted their appearance to mimic that of a more poisonous relative. In this way, they take advantage of markings that bring back unpleasant memories for predators. Such mimicry may protect the brightly colored, red-backed salamander from would-be predators, even though it is not toxic like the similarly colored eastern newt. Some snakes also mimic their poisonous relatives as a means of defense. The nonpoisonous scarlet kingsnake looks remarkably like the venomous coralsnake, both of which are found in the same region.

Finally, many herps scare off potential predators with threatening postures or behaviors. Snapping turtles, when encountered on land, can be very aggressive, snapping their jaws and lunging. Probably the most notorious warning among herps is the very distinctive and chilling sound of a rattlesnake's tail. The mere suggestion of a nearby rattlesnake is enough to make most animals halt in their tracks and make a hasty retreat. Some snakes will rise up as if poised to strike an attacker. This act also has the advantage of making them appear larger and perhaps more threatening. The hog-nosed snake, a common resident of the coastal plain, uses a complicated set of behaviors when it is attacked. It first elevates its head and spreads out the skin of its neck in an effort to look bigger and more threatening. If this doesn't scare off a predator, the hog-nosed snake begins to writhe upside down. It then regurgitates a foul smelling liquid and finally becomes rigid. It holds this position for several minutes, until the predator becomes disinterested and moves off.

Activity 3.1. Mimics Survive

The milksnake, kingsnake, and coralsnake have similar coloration. The harmless kingsnake and milksnake appear to be mimicking the patterns and colors of the poisonous coralsnakes.

Objective

Students will understand how successful the mimicry strategy is for confusing predators...in this case, them!

Materials

- drawings of Mexican milksnake and coralsnake (page 29)
- teacher guide to coloring Mexican milksnake and coralsnake (page 30)

Procedure

1. Cut out the figures of the two snakes, color them in as indicated, and hand them out to groups of three to four students.
2. Have students examine both figures independently and talk about the identifying characteristics (colors and patterns). Point out that the coralsnake can defend itself with a poisonous bite. Have students put the figures out of sight until Step 4.
3. Discuss how bright and distinct coloration is used to warn off predators from toxic prey. Give other examples of animals that use warning colors, such as the eastern newt and monarch butterfly.
4. Now have the students mix the figures up and try to identify which figure is which species. Can they tell them apart? Would you go near any of these snakes in the wild? What advantage does the Mexican milksnake gain from mimicking the coralsnake?

Activity 3.2. Don't Mess with Me

Objective

Students will understand a range of defensive strategies used by herps.

Materials

- live specimen or photo of the eft stage of the eastern newt
- dried rattlesnake rattles (available through biological supply companies)
- hand lenses

Procedure

1. If possible, have the students collect an eastern newt from the floor of the forest during the spring, summer, or fall. They usually can be seen walking around on

the surface by day, especially under moist conditions. Keep the newt in a moist environment, with leaves and soil for cover, and return it to its home within an hour or two at most. If a live specimen is not available, use the drawings found in the field guides listed on page 60.

2. Have the students examine the live specimen or photo of the eft stage of the eastern newt, noting the pattern of its coloration. Where is it brightest, back or belly? Drawing on the Eastern Newt Species Account on page 143, discuss the newt's life stages and migration and the fact that it is toxic.

3. Have students examine the rattlesnake rattle using a hand lens to get a close look. What is the rattle made of? How is it attached?

4. Have students compare how the red eft and rattlesnake warn off predators. Point out that the eft uses coloration to warn away potential predators and reduce the likelihood of being eaten while it moves between habitats. In contrast, the rattlesnake generally stays put and uses its rattle to warn away intruders.

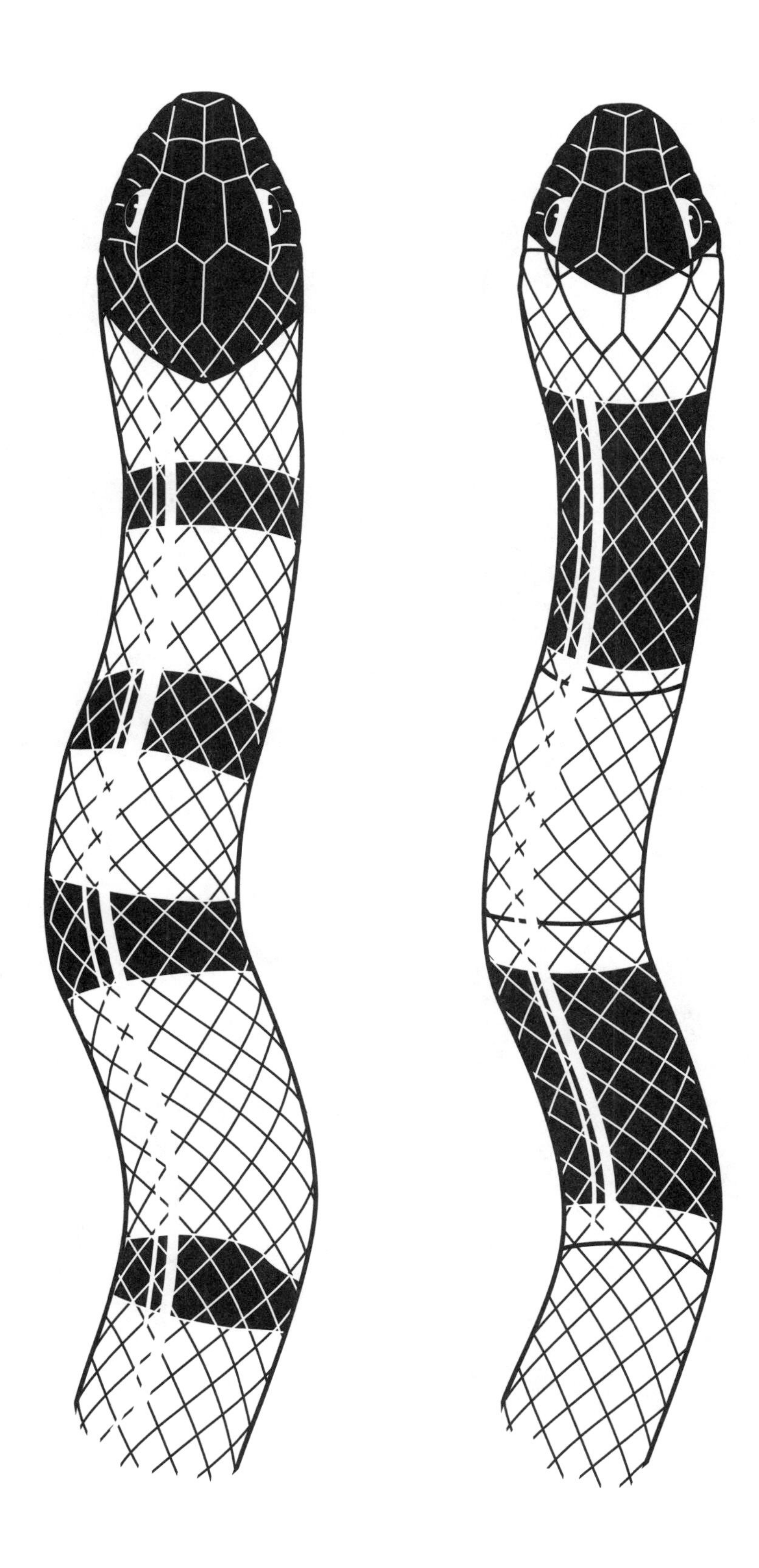

Teacher Guide to Coloring Mexican Milksnake and Coralsnake

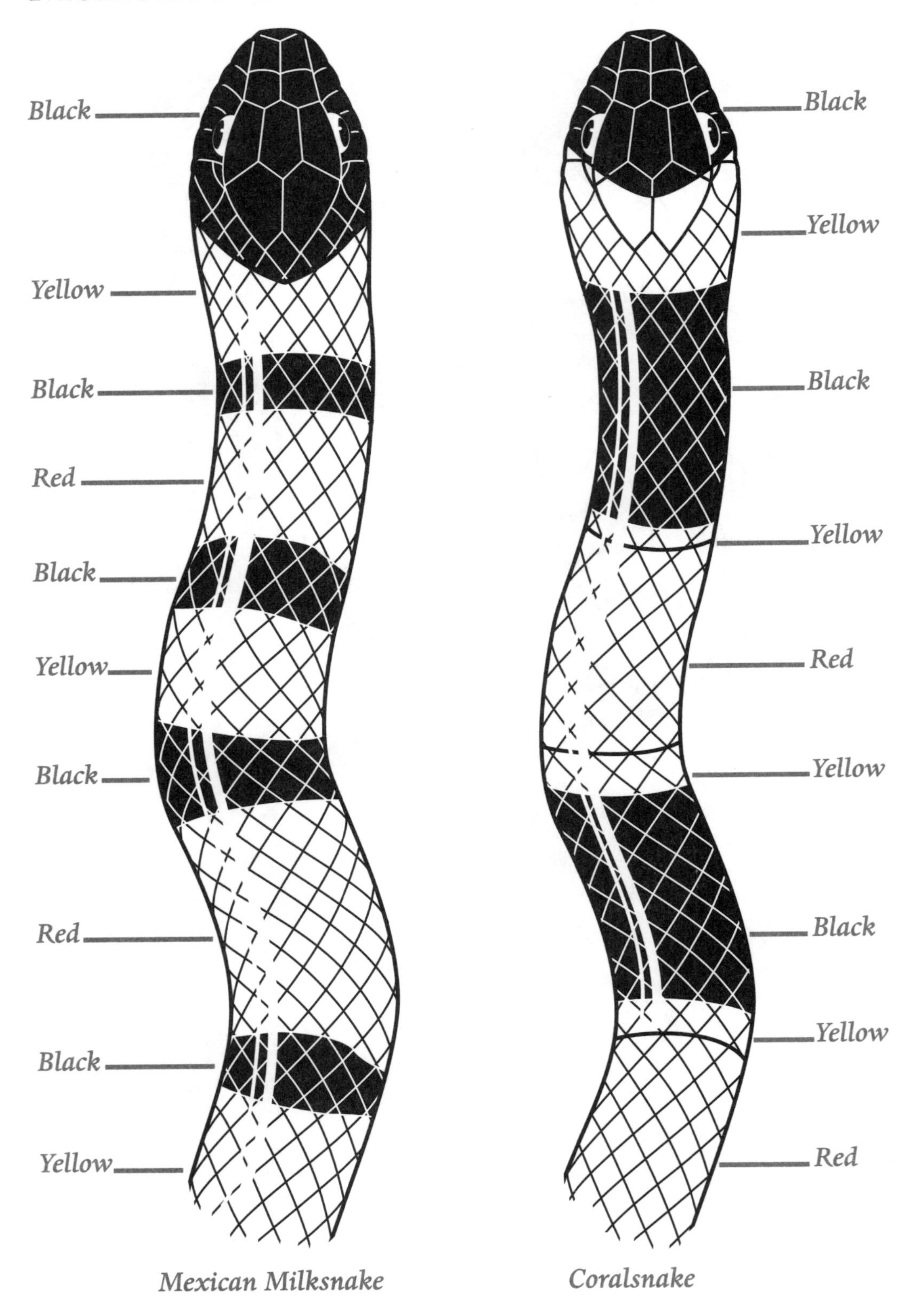

Chapter 4

Multiplying Herps—Reproduction and Development

National Science Education Standards
Grades 5–8: Reproduction and heredity (Life Science)
Grades 9–12: The behavior of organisms (Life Science

Although both amphibians and reptiles hatch from eggs, there are important differences between their eggs that greatly affect their habits and reproductive behaviors. Amphibian eggs are covered with soft jelly, rather than a tough outer membrane or shell. Because the jelly-covered eggs are very susceptible to drying out, amphibians need to lay eggs in moist or aquatic environments. In contrast, reptile eggs have extra membranes and an outer leathery shell that surround the developing embryo. These extra protective layers prevent the embryo from drying out. Thus, reptiles can lay their eggs in a much wider variety of habitats, even in dry and sandy places.

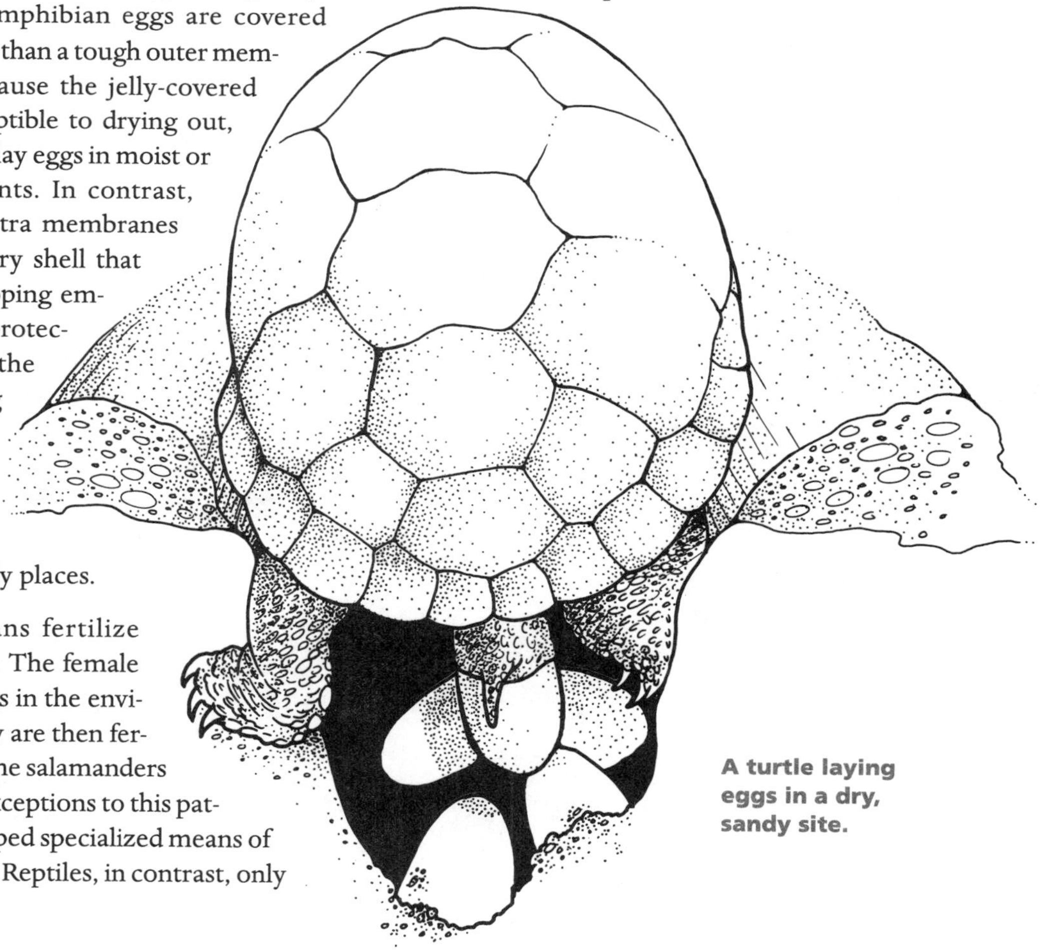

Most amphibians fertilize their eggs externally. The female first deposits the eggs in the environment, where they are then fertilized by a male. Some salamanders and a few frogs are exceptions to this pattern and have developed specialized means of internal fertilization. Reptiles, in contrast, only

A turtle laying eggs in a dry, sandy site.

fertilize their eggs while they are still inside the female. In this way, the fertilized embryo can be wrapped in its protective outer shell before the egg is laid.

Another major difference between amphibians and reptiles is the amount of development that occurs within the egg. Reptiles develop more completely within the egg and hatch out as small versions of the adults. They then continue to grow in size for most of the rest of their life with only slight changes in shape. Most amphibians, however, develop only minimally within the egg, and hatch out as larvae that are distinctly different from the adults. As they grow after hatching, they change dramatically in abrupt stages, a process called *metamorphosis*. The name *amphibian*, which means "double [*amphi*] life [*bios*]," refers to these drastic changes in body form and in the common transition from water to land.

Often, for amphibians that undergo metamorphosis, the first stage is spent as an aquatic tadpole. Tadpoles have tails to propel them underwater, and gills with which

Life Cycle of a Salamander

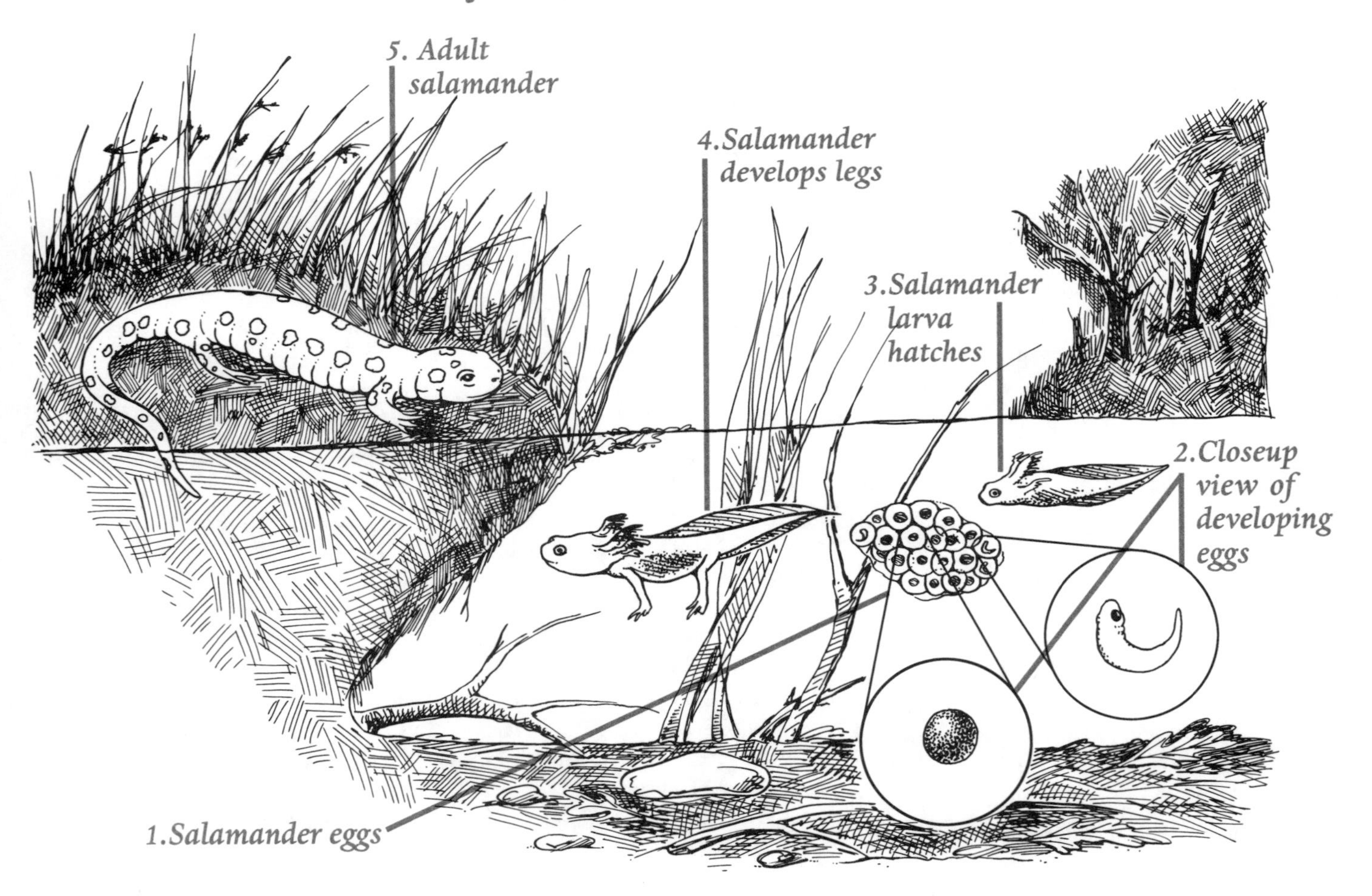

they breathe. Later, as they feed and grow, the tadpoles lose these aquatic structures, grow limbs, and develop lungs. Once changed, the amphibians are free to take up a new life on land. Some move out and remain near the water's edge, while other species travel great distances to live in surrounding fields and woods.

As might be expected, amphibians exhibit a wide variety of strategies in their development from eggs to adults. Among the salamanders alone, there are some great examples of this diversity. Some have distinct aquatic tadpole stages, with gills and tails, before undergoing metamorphosis and moving onto land. Others develop much more within the egg, and differ little from the adults, except for the small gills they use to breathe in water. In some species, such as the mudpuppy, the juveniles never lose their gills, retaining them even as adults. By not changing their body form, they can remain in an aquatic environment throughout their entire lives. Still other salamanders live their entire lives on land, from the time when they first hatch under rocks in the forest.

Reptiles lay their eggs in a variety of habitats, but always on land. For many reptiles that deposit their eggs in nests, an interesting process occurs during incubation. In most turtles and crocodilians, and in some snakes and lizards, the nest temperature is the main factor in determining the sex of the egg. In turtles, nest temperatures above 30° C (86° F) produce females, and lower temperatures yield males. The reverse is true for alligators and other herps, where higher temperatures produce males. Some of the larger reptiles such as seaturtles place nests of more than 100 eggs deep beneath the sand. Under sunny conditions the upper parts of the nest can be a little warmer than the bottom, creating female hatchlings near the top and males near the bottom. So the difference between being a male or a female may depend on which way the egg rolls when it drops into the nest.

This process, called temperature-dependent sex determination, may seem radical, but some scientists believe that this was the original way of determining sex among all ancestral reptiles. No matter the origin, it is a common concern that global climate changes leading to steadily increasing temperatures will have the devastating effect of driving these reptiles to extinction.

Some amphibians and reptiles have evolved the ultimate protection of retaining their eggs within their bodies until they hatch. The process of "livebearing" is achieved by different groups of herps in different ways. In some species the eggs simply develop and hatch within the parent, receiving little or no nutrients throughout the incubation period. In others, such as gartersnakes, rattlesnakes, and many lizards, the mother's body provides nourishment for the developing embryos during pregnancy, similar to what occurs in mammals.

In most cases where eggs are deposited externally, they are left to develop on their own without further care by parents. However, several amphibians, and a few reptiles, provide protection to their eggs during development. This parental care can have a big influence on the survivorship of offspring. Some newly hatched salamanders even remain near their mother for several months, taking advantage of the shelter and scraps of food provided. Similarly, young alligators often spend their first years under the protective umbrella of one or both parents. When a young alligator yelps in distress, anyone with experience knows that it's time to leave fast, before the person is confronted with a furious mother or a pond full of angry relatives. This protective behavior is exceptional among the herps, since most of them spend little or no energy providing for their young. However, the wide variety of behaviors among the species is merely reflective of the incredible diversity these animals exhibit in reproductive strategies, as well as many other key biological functions.

Activity 4.1. Amphibian Life Cycles

This activity must be performed in spring and summer when there are eggs of frogs and salamanders in ponds and ditches.

Objectives

Students will (a) observe the development and metamorphosis of amphibian eggs, and (b) gain an understanding of the different stages (or steps) of amphibian development.

Materials

- clusters or strings of amphibian eggs from a nearby pond
- aquarium
- buckets, pans, and/or plastic containers with lids to collect water, rocks, and aquatic plants
- plastic containers to collect amphibian eggs
- pond water
- thermometer
- hand lens or dissecting microscope
- logbook(s) to record observations

Procedure

Safety Tip: Be sure students are dressed appropriately for getting wet. Supervise young children around the pond edge. Consider having youth wear life jackets if the pond is deep or steep-sided.

1. Set up the empty aquarium where your group can watch the eggs hatch and grow. Do not put the aquarium in a sunny window or other location where the water is likely to become too warm.
2. Locate a pond in which amphibian eggs are present.
3. Assemble all your collecting equipment, including containers to collect eggs, buckets or plastic jugs to collect pond water, thermometers, hand lenses, and log books.

4. Take your students on a trip to the pond. Have them collect enough pond water to fill the aquarium at least three-quarters full and enough rocks to fill a corner of the aquarium above water level. Slimy rocks are good; they will provide algae for food. Fill a bucket or pan with small aquatic plants and enough of the soil they are growing in to anchor them to the bottom of the aquarium.
5. If you haven't already found them, look around the pond for floating egg masses and strings or globs of eggs attached to the plants in the water along the edge of the pond. Each egg will be about 5–8 mm (.2–.3 in) in diameter, clear with a black dot in the center. Wait to collect the eggs until after the next step.
6. Now is the time for the students to record their observations of the site. What is it like? Shady, sunny? What temperature is the water? Where did you find the eggs you are collecting—on the bottom, the surface, free-floating, or attached to plants? How abundant are these eggs in the pond? Are there signs of other animals nearby—for example, tracks, broken branches, scat?
7. Most frog egg masses will have hundreds of eggs. Toad eggs will occur in long strips. Salamander egg masses tend to have 50 or fewer eggs, and the individual eggs are larger than frog and toad eggs. Frog and toad eggs are preferable for this activity. You only need to collect 30–40 eggs. Be careful not to destroy the entire mass of eggs in the pond. Have students carefully collect the eggs by separating them gently and letting them float into a jar or plastic bag. Allow room for air in the top, close the container securely, and return home or to the classroom.
8. Have students place a layer of soil in the bottom of the aquarium. Stack the rocks in one corner. Fill with pond water. Secure plants in the soil. After the water has cleared, add the egg masses. Count or estimate the number of eggs at this time and record in the logbook.
9. Now you can watch the wonder of metamorphosis. The eggs should hatch in two or more weeks. The animals will develop tails while still in the egg. Salamanders will also develop legs. Because the eggs are transparent, you can observe embryo development by putting a few eggs in a water-filled dish and looking at them with a hand lens or dissecting microscope. Keep these eggs in tank water and do not let them heat up. Return them to the tank as soon as possible.
10. Have students record their observations in the log at regular intervals. They should keep track of how many eggs hatch and how many survive to become frogs or salamanders.
11. You may want to have the students graph the number of eggs, number of individuals that hatched, and number of individuals that survived through metamorphosis. They could also graph the number of eggs hatched each day and number of tadpoles that metamorphosed each day. Lead them in a discussion of why many more eggs are laid than will eventually become adults.
12. Maintenance is a critical part of this activity. Twice each week, replace the aquarium water with fresh pond water. Once the tadpoles have hatched, they

will feed on algae on the rocks and plants and tiny zooplankton in the water. Keep the water cool and out of bright sunlight.

13. Once the animals have metamorphosed, return them to the pond where you found the eggs. They will not continue to live in the water after they have changed.

Chapter 5

Breeding Behavior

National Science Education Standards
Grades 5–8: Reproduction and heredity (Life Science)
Grades 9–12: Behavior of organisms (Life Science)

The evening chorus of frogs is a welcome sign of spring throughout much of the United States. You can easily hear spring choruses as you pass by ponds, marshes, creeks, roadside ditches, and even puddles.

Unlike other amphibians and reptiles, most frog species call out to attract mates. To amplify the calls, frogs fill one or two balloonlike sacs on their throat. They then force out air from their lungs past their vocal chords, causing the chords to vibrate. The expanded sacs act as resonators, amplifying the sound much in the same way as the hollow space in a drum. Sometimes when hundreds of frogs begin calling at once, the sound can be downright deafening.

Spring peeper with throat extended to amplify mating call.

The main purpose of calling out loud is to attract mates. In general, it is the males that sit in one place and call to attract females. The female often is drawn to the male who calls the loudest or most frequently. In some species, the female chooses a male with the deepest voice, whereas others select their mate by the quality of his calling site. Once the female accepts the male, he embraces her from above, squeezing her sides with his forearms. She then squeezes out her eggs into the water, where they are fertilized by the male's sperm. Frogs in the United States tend to lay their eggs in clumps or singly, attached to sticks and vegetation near the surface of the water. Eggs that are in continuous strings usually are from toads.

Many different species may be calling in the same area during the same brief period of time each year. Therefore, it is critical that the females recognize their own species. This usually is accomplished by each species having a distinct call, unique in its pattern and pitch. In some species, the female's ear can hear only in the wavelengths of the male's calls.

Calling for mates is a great way for a frog to advertise its presence and willingness to mate. However, making loud noises also can attract unwanted attention from predators. Being active under the cover of darkness partially reduces the chances of being located. Nevertheless, a calling male is still a vulnerable target for raccoons, fish, nocturnal birds, and some bats. Calling together in large groups provides some protection for the frogs and toads. Predators find it difficult to focus on a single toad in all the confusion of noise that surrounds a chorus. This principle of "safety in numbers" is also one of the reasons that birds flock, fish school, and mammals herd together.

Calling out loud is only one way to attract mates. Many other amphibians and reptiles also congregate during the mating season, but do it silently. Each year at the end of winter, as the ground thaws, hundreds of spotted salamanders suddenly appear in woodland ponds throughout the eastern United States. They seem to appear from nowhere all within a few days. But, instead of calling, the spotted salamanders communicate with each other by behavioral and chemical signals. They mate in a silent flurry of activity and quickly retreat to the obscurity of the woods until the next year.

Activity 5.1. Visit an Evening Chorus

Objective

Students will understand the importance of calling as part of the breeding behavior of frogs and toads.

Materials

- Tape or CD of frog calls (optional; see list below)

Procedure

Safety Tip: Be sure students are dressed appropriately for getting wet. Supervise young children around the pond edge. Consider having youth wear life jackets if the pond is deep or steep-sided.

1. Locate a nearby pond or wetland. Check for signs of frog and toad breeding in early spring (March or April) when the daytime temperatures average above freezing. You should first hear a few frog calls at twilight or during an overcast day. Within a few days, most of the frogs in that population begin calling, and the chorus will be at its peak. This chorus will last for about two weeks. It ends when breeding is over and the eggs are laid in the pond. If you see eggs afterward, you can match these up with the species of frogs that you observed.

2. Have the students listen to a tape or CD of frog calls and become familiar with the calls of local amphibians.

3. Visit the pond in early twilight. You must be very quiet as you move close to the pond and keep still while listening. Also, if you visit the pond during peak migration of frogs and salamanders, take care not to drive over animals crossing the road while moving between upland and pond sites. Warn the students to take care not to step on animals either. Because the number of animals can be overwhelming, it is often extremely difficult not to crush them. You may want to park away from the site and walk slowly using a flashlight. Try to minimize the number of different nights you go out and disturb one site.

4. Use a flashlight at intervals to try and see the frogs that are calling. They usually will be sitting along the water's edge, floating in the water, hanging on stems of wetland plants, and sitting on floating leaves.

5. Have students describe the different types of calls. Have them estimate how many species are present and how many animals are calling.

Frog Call Tapes and CDs

Bogart, C. M. *Sounds of North American Frogs: The Biological Significance of Voices in Frogs.* Smithsonian Institution. Folkways Cassette Series 06166. Center for Folklife, Programs, and Cultural Studies. 955 L'Enfant Plaza, Washington, DC 20560. 800-410-9815.

Elliot, L. 1992. *The Calls of Frogs and Toads: Eastern and Central North America.* Post Mills, VT: Chelsea Green Publishing.

Kellogg, P. P., A. A. Allen, and T. Wiewandt. 1982. *Voices of the Night: The Calls of Frogs and Toads of Eastern North America.* Cornell Laboratory of Ornithology. 159 Sapsucker Woods Rd. Ithaca, NY 14850.

McGrath, J., and A. Clay. *Frogs of the Lower Great Lakes Region.* Nature Discovery. 5900 N. Williamstown Rd. Williamstown, MI 48895. 517-655-5349. *www.lansinggrain.com/frog/*

(For a complete list, see *monitoring2.er.usgs.gov/FrogWatch/How/Learn/Vocalizations/vocalizations.htm*)

Chapter 6

Life Span and Life History

National Science Education Standards
Grades 5–8: Populations and ecosystems (Life Science)
Abilities necessary to do scientific inquiry (Science as Inquiry)
Grades 9–12: Abilities necessary to do scientific inquiry (Science as Inquiry)
Population growth (Science in Personal and Social Perspectives)

How long do reptiles and amphibians live? Some species of lizards and frogs live short lives of only one or two years. It is relatively easy to study the life stages and longevity of these short-lived species. However, for long-lived species, much less is known about life span and age-related processes. Studying these animals requires monitoring individuals in the wild over many years. Most often this is expensive and hard to do. Many herps outlive the life of the study, and some even outlive the researchers. Therefore, much of the information on animal life spans comes from zoos that keep careful records.

Some of the largest turtles, such as giant tortoises, may survive for as long as 100 years. But size is not always related to life span. The small Blanding's turtle can live for more than 60 years and a common ring-necked snake that you encounter can be 15 years or older. Even some of the smallest salamanders can live for 10 to 15 years or more.

How can you tell the age of an animal in the wild? In the case of a turtle, its bony shell is covered with a series of dry scales called scutes. As the turtle grows, it adds a new layer underneath the old scute, only bigger. This results in layer upon layer of scutes. Some turtles shed outer scutes periodically, while other turtles retain their outer scutes until they are worn off by years of abrasion. In either case, an interesting phenomenon occurs with turtles in colder climates. The turtles grow only during the warm weather. When the turtle's growth slows down in the fall, it forms a distinct line around the edge of the scute. As new scutes grow underneath, a remnant of past lines remains. Because one new line is added each winter, you can count the lines to get an estimate of the age of the turtle. This method does not work for soft-shelled turtles, seaturtles, and some other species. But if you look at a single scute of a painted turtle, snapping turtle, or diamond-backed terrapin, you can get a fairly good idea of the turtle's age.

Because it is possible to determine the age of turtles, they make ideal subjects for studying age-related biological processes. Scientists often want to know the number of individuals of different ages in a population of turtles. This allows them to determine whether the population—the number of turtles in a given area—is growing or declining. However, it is difficult to locate and age every turtle in a wild population. Therefore, the scientists carefully select a random sample of individuals in a population and then use statistics to analyze the data and to describe the larger populations.

When studying a population of turtles, scientists measure and weigh each animal sampled. When appropriate, they also count scute rings to estimate the turtle's age. They record the time and location of capture and often label each turtle on its plastron (lower shell) or carapace (upper shell). A very clever method of keeping track of individuals in a population is to photocopy their plastrons. Because each individual is unique, this is as effective as taking fingerprints of humans. Each time an animal is captured and measured throughout its life, a lot of important information can be learned about its biology, such as movement, growth rate, and life span.

When studying wild animals, and especially when making decisions about conservation and management, knowing the life span is not as important as knowing the relationship between the age of an animal and important events in its life. In particular, the average age at which the members of a species begin to reproduce, the number of offspring produced by each mother, and how often they reproduce throughout their lifetimes are important measures of the species' ability to persist in nature. When births are few or far between, the importance of each individual increases. When a species takes a long time to reach maturity, the relative importance of adults in the population also increases. Animals that have low rates of reproduction often are vulnerable to environmental disruptions and thus we should put greater effort into protecting individuals.

Many reptiles and amphibians take years just to reach sexual maturity. For example, a painted turtle may take five years to reach sexual maturity whereas a snapping turtle may remain immature for its first 10 to 15 years. A few decades ago, the American alligator was nearly driven to extinction even though only the older, larger animals were being hunted. What we found nearly too late was that no alligators were being allowed to get old enough to reproduce. Once we figured out how long it took an individual to reach maturity, we were able to successfully repopulate the American alligator across the southeastern United States. Thus, by understanding age-related processes such as life span and the timing of reproduction, we can design better strategies for conservation.

Activity 6.1. Long-Lived Turtles

In this activity, students learn to estimate turtle size and age. This activity builds on the handling and care concepts introduced in Activity 2.1. If you choose not to do 2.1 first, then begin by demonstrating the safe handling of turtles (see "Care and Handling of Live Herps," page 2). Also, be sure to have students wash their rulers and hands with antibacterial soap or wipe them with hygienic wipes. Point out the importance of doing this to avoid infection from the turtles and to avoid the possibility of spreading infection from one turtle to another.

Objective

Students will learn how to estimate the age of turtles and how to measure a turtle's size.

Materials

- live turtle(s) or turtle shells
- flexible transparent rulers with millimeter scale
- copies of two diagrams: Measuring Turtle Length and Estimating Turtle Age from Scutes (provided)
- hygienic wipes or antibacterial soap

Procedure

1. Demonstrate safe handling of a live turtle.
2. Identify the parts of the shell including carapace (back), plastron (underside), and scutes (scales).
3. Refer to diagram to show students how to measure the length of the turtle's plastron along the midline from one end to the other. Then demonstrate how to measure the carapace length using first the straight-line method and then the curved-line method.
4. Have students measure the length of each turtle shell and record their measurements.
5. Point out the lines on the scutes of either the carapace or plastron. Use diagram to demonstrate how to count the lines on one scute. The number of lines on each scute represents an estimate of the age of the turtle.
6. Have the students count the lines on one scute and record their observation.
7. Ask students to estimate the approximate age of the turtle based on the number of lines on a scute. If you have several turtles or turtle shells, students can relate the age to the size of turtles. They may want to draw a graph showing age on the x-axis and size on the y-axis. Have the students discuss how big the turtle would be at sexual maturity if it occurred at age 5, age 10, or age 15.
8. Discuss factors that may influence the growth of turtles and formation of growth lines.

Activity 6.2. Estimating Madness

This activity is very similar to Activity 6.1 except that it adds the concepts of sampling error and variability in scientific data. It is appropriate for more advanced students who are learning about research and ways to analyze data.

Objectives

Students will (a) learn how to estimate age and size of turtles, and (b) understand variability in scientific data.

Materials

- live turtle or turtle shell
- flexible transparent ruler with millimeter scale
- copies of two diagrams: Measuring Turtle Length and Estimating Turtle Age from Scutes (provided)
- small sheets of paper on which to record measurements
- calculator
- blackboard with chalk
- hygienic wipes or antibacterial soap

Procedure

1. Select one turtle or turtle shell and follow steps 1 through 4 in Activity 6.1 (Long-Lived Turtles).
2. Have each student independently measure the turtle's carapace length and record the measurement on a piece of paper. Be sure that everyone uses the same method, either straight-line or curved. Also have each student count the number of lines on a scute and record the estimate of the turtle's age.
3. Assign two people the job of reading the measurements off the slips of paper and recording them in tables on the board. Did everyone use the same units of measure? If not, convert the values to the same units. Calculate the average of the measurements for the length and age of the turtle.
4. Identify the maximum and minimum values reported for length and age of the turtle. These values will give the students an idea of the variability involved in measuring. Have the students talk about exactly how each person measured the shell and compare techniques. The variability in how they measured the shell is a type of sampling error. It always occurs in sampling and can be a problem for scientists trying to gather exact information. Some of the sampling error is due to mistakes, whereas some is due to differences in researcher technique. Discuss with the students how they might reduce this sort of variability. For example, teachers might provide clearer directions for taking measurements or researchers might practice on a model or separate animal first.
5. Ask the students to identify other kinds of variability that might affect the accuracy of their results. For example, turtles may not add exactly one growth line per year, so counting the lines may not give an accurate measure of age. Also, as turtles get older, their shells often get so worn that some or all of the growth lines may be obliterated.

6. Have the students take the entire range of sizes recorded and divide it into five equal intervals. Draw a graph with the size intervals on the x-axis and numbers on the y-axis ranging from zero to the total number of students. Have the students plot the total number of measurements that were recorded by them within each of the size intervals. This is a graph of the distribution of the variability of the measurements. If all the measurements were the same, there would be only one point on the graph. Usually, however, you will see a bell-shaped curve, with the most measurements occurring at the average size and a few measurements near the minimum and maximum sizes. This is the famous bell curve of a normal distribution. In general, the more measurements you have, the smoother the curve.

7. Finish by talking about why there might be more measurements near the average size. Much of the statistical analysis that researchers perform begins by examining the information in the same way that the students have just done.

Safety Tip: Have the students wash their rulers and hands with antibacterial soap or wipe them with hygienic wipes. Point out the importance of doing this to avoid infection from the turtles and to avoid potentially spreading infection from one turtle to another.

Measuring Turtle Length

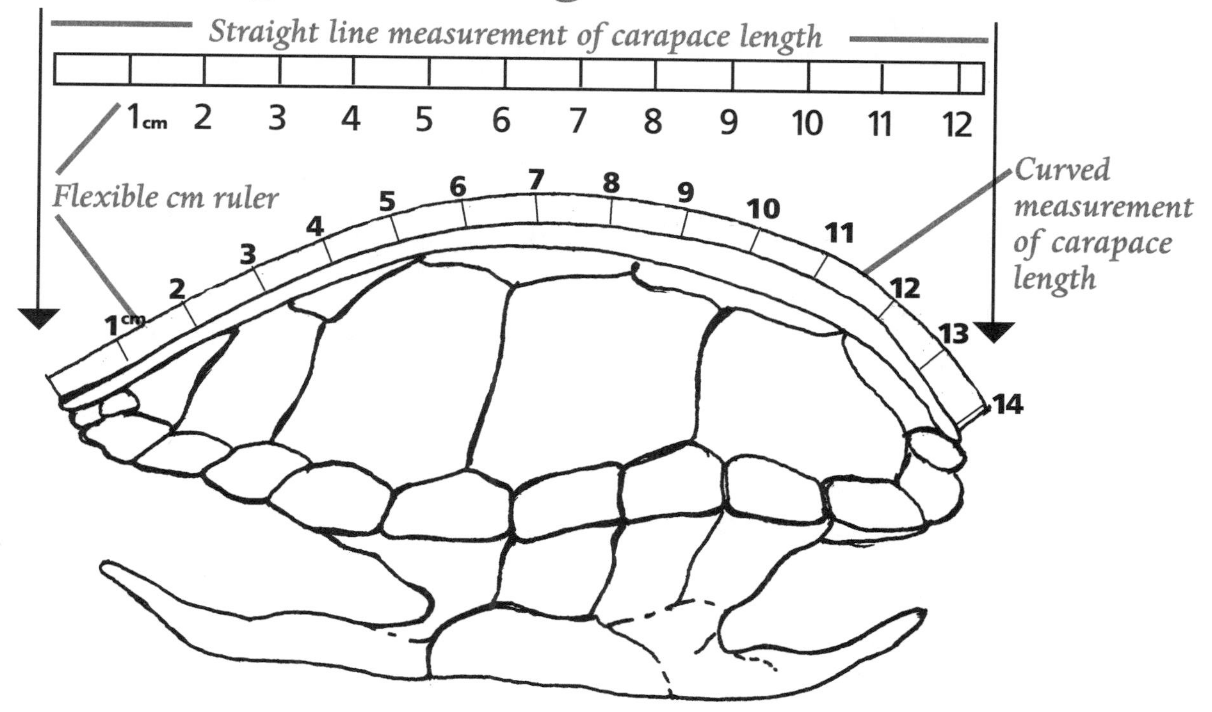

Estimating Turtle Age from Scutes

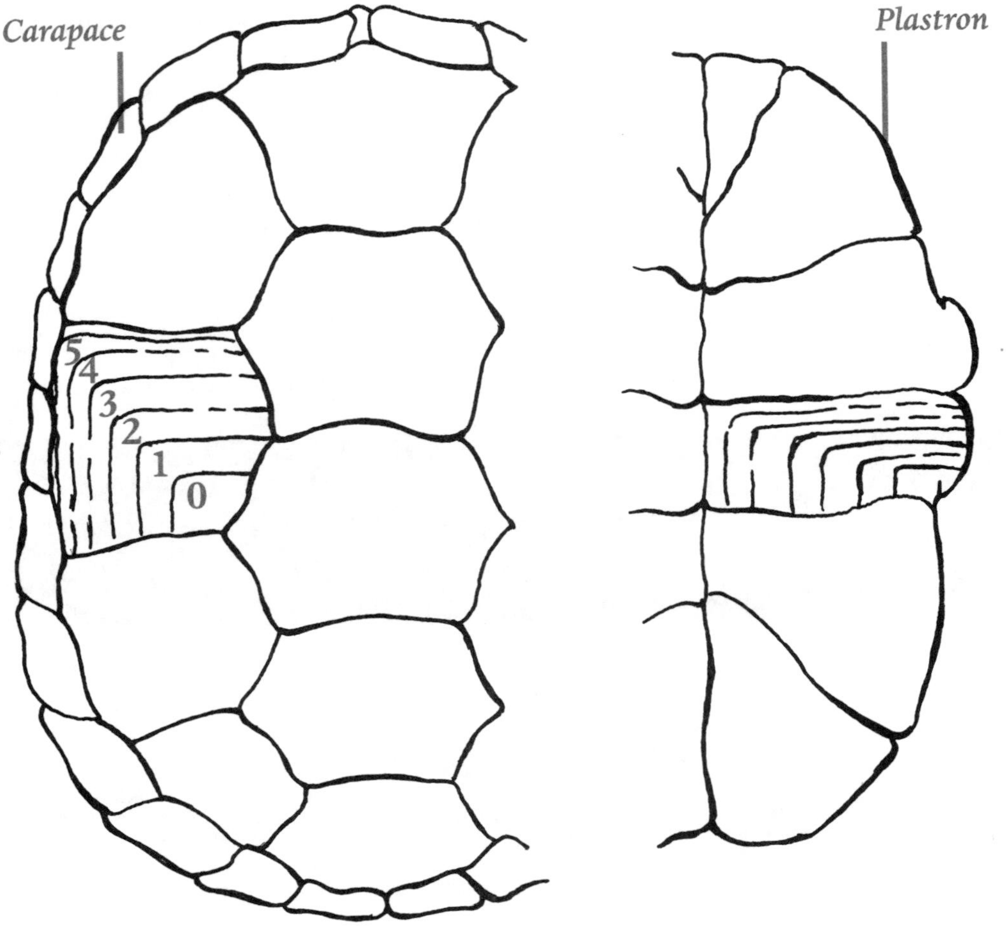

Carapace

Plastron

(The number 0 indicates the scute that is present at hatching; 1–5 indicate scutes formed during five subsequent growing seasons.)

5
4
3
2
1
0

Activity 6.3. Turtle Shell and Egg Hunt

This activity takes place during late summer or early fall.

Objective

Students will begin to apply life history concepts to a natural population of turtles.

Materials

- pond or wetland known to have a population of turtles
- flexible transparent ruler
- notebook and pencil
- bag to hold turtle shells

Procedure

1. Scout out several different ponds or wetlands ahead of time to locate one that supports a population of turtles. The quickest way is to stealthily approach and quietly observe the pond during a warm spring day to see if turtles are basking on logs or along the pond shoreline.

2. Take the students to the pond. Have them walk carefully around the perimeter of the pond looking for empty turtle shells or nest sites with piles of eggshell remnants on the ground surface. The eggshells will look like pieces of Ping-Pong balls or curled up white paper. Nests generally will be located within 200 m (50 yd) from the water's edge.

3. Have the students collect any shells of dead turtles that are found, and record the location, distance from the pond, and other habitat information. Have students estimate the number of eggshells found at nest sites and record information about the associated habitat, including distance from pond, soil type, and nearby vegetation. Note any evidence of the nest being dug up by predators.

4. Back at the classroom, have the students measure the size and estimate the age of each turtle shell (see Activities 6.1 and 6.2).

5. Have the students combine and discuss their information from the nests. How far were the nests located from the pond? Was there any consistency in the habitat associated with all the nests? Was the number of eggs found in each nest the same? What does the number of nests suggest about the number of females that may be present in the pond?

6. Have the students summarize and discuss the information collected from the turtle shells. What is the average size and age of the turtle shells collected? Make some guesses as to whether these represent the older or younger turtles in the population. Which turtles would the students expect to find on land and why? Discuss any evidence, such as a crushed shell or animal tracks near the site, that might indicate how each turtle died.

Chapter 7

Herps through the Ages—Evolution and Extinction

National Science Education Standards
Grades 5–8: Diversity and adaptations of organisms (Life Science)
Grades 9–12: Biological evolution (Life Science)
Natural and human-induced hazards (Science in Personal and Social Perspectives)

The earliest amphibians appear in fossils that are over 370 million years old, whereas reptiles date back around 320 million years. Interestingly, over a geologic time scale, reptiles such as crocodilians and turtles have not changed very much in their appearance or habits. These ancient animals wandered the planet with the dinosaurs and have survived millions of years after the last dinosaurs disappeared.

Many other amphibians and reptiles, however, have undergone dramatic changes over time. The earliest amphibians looked more like fish with legs. From these early semiaquatic creatures evolved a wide variety of life forms that barely resembled their primitive ancestors. Over hundreds of millions of years, many different amphibians evolved and disappeared, leaving the familiar salamanders and frogs of today.

The early amphibians were such a successful group of animals that the basic four-legged body plan continued to prosper and evolve into many different forms. The first radically different animals to arise from this lineage were the reptiles, which quickly spread to become the dominant life form on land. Over tens of millions of years, these animals flourished, taking on all sorts of shapes and modes of life. Included in this group were the huge dinosaurs, flying and swimming reptiles, and present-day snakes and lizards. From that one original amphibian lineage also stemmed modern-day birds and mammals.

This intricate and ancient history of evolution and family relationships among amphibians and reptiles has been painstakingly reconstructed by untold number of paleontologists working around the world for several hundred years. Paleontologists are like detectives who piece together the evolutionary record using fossils of organisms that existed long ago. They examine closely the shape, size, and arrangement of bones in fossil skeletons, as well as the position and abundance of the bones compared to the age and other features of the surrounding rock. Using these fragmented records, paleon-

tologists are able to theorize about the ancestry and relationships of species and rates of evolutionary change and extinction.

Why have crocodilians changed so little from their ancestors while other groups have changed so much? Paleontologists and biologists believe that millions of years ago these animals developed survival strategies that have been successful despite many episodes of extreme changes in the environment. Crocodilians are generalists rather than specialists. With strong jaws, sharp teeth, and thick skin, they are top-level predators in their food web, preying on many different animals both on land and in water. But it is not beneath them to eat something that is already dead or rotten. They also have an incredible ability to cope with variable and sometimes harsh environments. They can go prolonged periods without food, they can withstand a wide range of temperatures, and many can live in freshwater and saltwater environments. If a wet area dries up, an alligator or crocodile can walk great distances over land to another pond. In this way, it is believed that crocodilians have hung on, while many other species disappeared.

In some instances in the past, environmental conditions appear to have changed so much that only the most resilient animals persisted. Paleontologists have demonstrated that the global fossil record tells of a great catastrophe around 65 million years ago. Dinosaurs, as well as many invertebrates and other organisms, disappeared at an astonishing rate at this time. What happened? Scientists have proposed many different theories to explain the abrupt disappearances. Recent explanations suggest that either an asteroid collision with the Earth, or global volcanic eruptions, forced clouds of debris and dust into the atmosphere. The subsequent blocking of the sun's warming light resulted in cooling of the global air temperatures by several degrees. Apparently the many species that died off were incapable of withstanding such major changes in their environment. Many of these organisms had become highly specialized to certain modes of life and were not able to adapt to the new conditions.

It is clear when looking at animals of today, or in the fossil records from millions of years ago, that a wide variety of life forms and a vast array of strategies can be successful. Some strategies work well under constant environmental conditions, whereas other strategies work well under variable or changing conditions. Animals that are specialists often prosper when conditions to which they have adapted remain unchanged. They often have evolved features that allow them to take advantage of unusual foods or very specialized habitats. For example, spring salamanders survive well in the highly specialized environments of hillside springs and creeks along steep slopes, where conditions are dark, wet, and cold and few other vertebrates survive. However, specialists often are threatened when their unique resources are taken away. Spring salamanders generally do not tolerate increases in water temperature, reduction in oxygen, or the introduction of predatory fish. In contrast, generalists such as crocodiles and turtles tend to continue along during times of change, withstanding and persisting.

Within this framework of evolution, adaptation, and environmental change it is expected that the extinction of certain species would be inevitable over the immense scale of time that life has existed on Earth. Thus, extinction would appear to be a natural process. However, in the past 40 years scientists and nature enthusiasts have witnessed a dramatic decline in many species of amphibians and reptiles. In order to sort out the normally occurring process of extinction from some modern-day disturbances, we must draw once again on the expertise of paleontologists. By reconstructing the past history of plants and animals through the fossil record, they have determined that the rates of extinction we are experiencing today far exceed those that have occurred before humans dominated the Earth. The causes for global declines of amphibians and reptiles are complex, but there is little doubt that much of the trouble is caused by disturbances to the environment brought about by an ever-growing human population.

Activity 7.1. Dinosaur Detectives

Objectives

Students will learn about amphibian evolution by comparing fossils with a modern-day skeleton. They also will experience the scientific thought process of drawing conclusions from limited paleontological data.

Materials

- photocopies of the four figures of skeletons (provided)
- photocopies of the geologic time scale (provided)
- scissors

Procedure

1. Photocopy the four figures of herp skeletons and geologic time scale so that each student or team has one copy of each.
2. Have the students observe the four figures carefully. Then have each team use its best judgments to order the figures from most ancestral to most recent animal. Remind them that early amphibians evolved from a fish ancestor.
3. Discuss the correct sequence of evolutionary development among the four organisms. The correct order is as follows:

   B. *Icthyostega*. One of the earliest fossil skeletons of a four-legged animal ever found. It represents the earliest transition from the fishlike ancestor to a land vertebrate with weight-bearing limbs. This organism lived during the Upper Devonian, 360 million years ago, in what is now Greenland and Russia. Note that this fossil was incomplete, missing its front toes.

   D. *Temnospondyls*. One of many predecessors to the modern amphibians that lived over the immense span of time between 335 and 210 million years ago.

   A. *Triadobratrachus massinoti*. This small animal (10 cm [4 in] long) lived in the Triassic period 230 million years ago. It bears the strongest resemblance to modern-day amphibians.

   C. Skeleton of a modern-day frog. Modern frogs do not appear in the fossil record much before 100 million years ago.

4. Have the students align the figures with the time on the Geologic Time Scale when the animals lived. Discuss some of the key changes that occurred along the evolutionary pathway that led to the form of the modern-day frog:

   a. loss of the tail

   b. fusion of the head bones

   c. reduction in the number of (digits) toes

   d. reduction in the ribs and vertebrae

   e. elongation of the digits

   f. elongation of the leg bones

5. Discuss how paleontologists use information from rare fossils collected at different times and from different places around the world. Using these bits of evidence, paleontologists piece together possible pathways in the evolution of amphibians and reptiles. Discuss the uncertainty involved in working with such limited information.

Geologic Time Scale

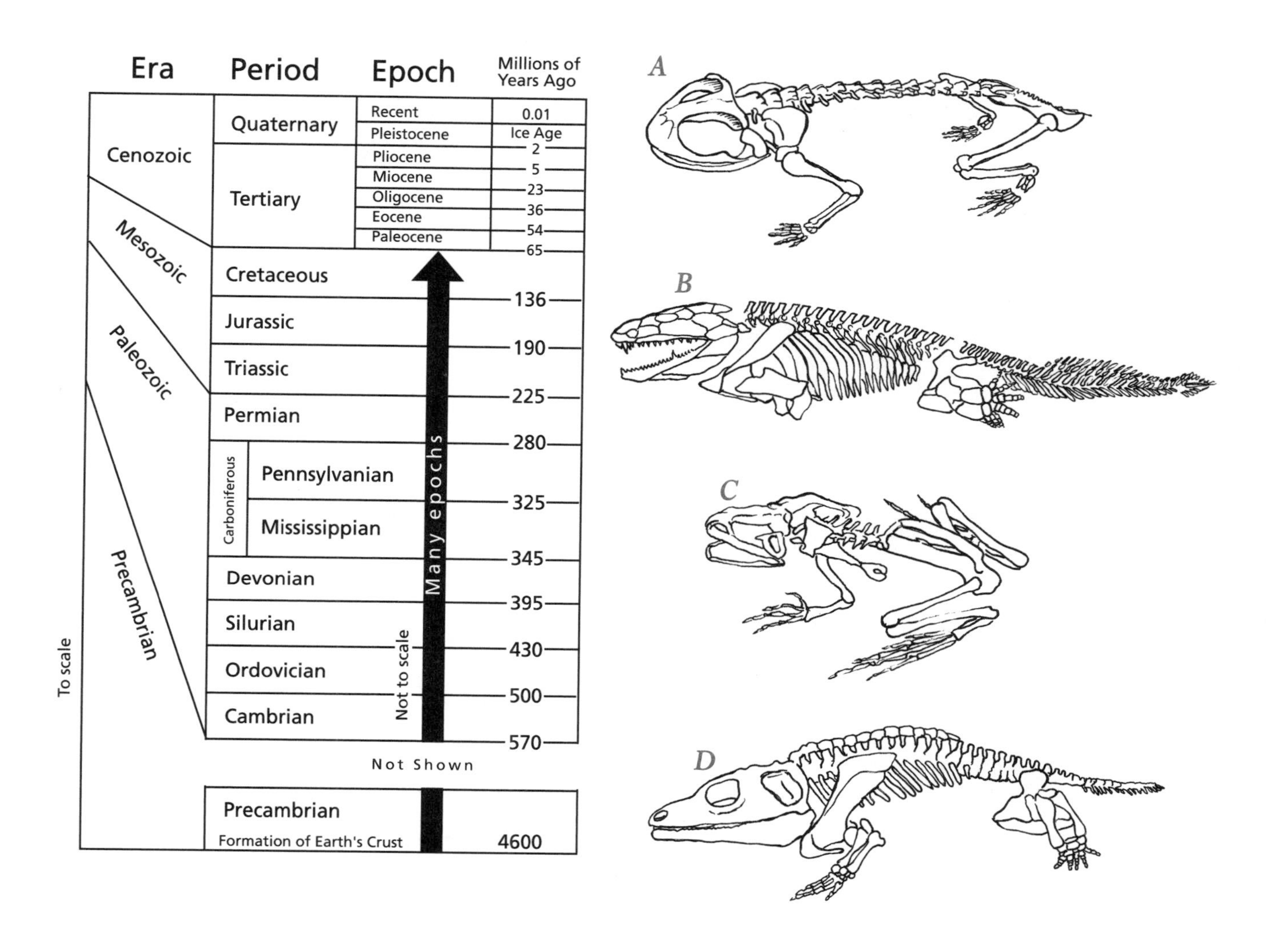

Source

Skeletons based on drawings in F. H. Pough *et al.* 1998. *Herpetology*. Upper Saddle River, NJ: Prentice Hall.

Herps: A Role in the Bigger Ecosystem

Amphibians and reptiles are integral components of their environment. But their generally small size and secretive behaviors often obscure the important role they play in the bigger ecosystems they inhabit. To truly appreciate herps, you need to experience the awesome variety of species and the huge number of individuals that occur throughout the landscape. The four chapters in this section are designed to provide opportunities for youth to interact with herps in their native habitats while learning about their diversity, abundance, relationships with other organisms, and habitat needs. Each of these activities can be modified for use with all ages of youth.

Chapter 8. Monitoring Species Diversity and Abundance

Chapter 9. Herps and Their Homes

Chapter 10. Food Webs

Chapter 11. Upland-Aquatic Linkages

Chapter 8

Monitoring Species Diversity and Abundance

National Science Education Standards
Grades 5–8: Diversity and adaptations of organisms (Life Science)
Abilities necessary to do scientific inquiry (Science as Inquiry)
Grades 9–12: Abilities necessary to do scientific inquiry (Science as Inquiry)

Each ecosystem ultimately is limited in the total number of animals and plants and the total number of different species that it can support. There are two general terms used to describe these measures of the animal and plant community: *abundance* and *species diversity*. Abundance is simply the number of individuals of a given species or a group of organisms in a given area. Species diversity is a measure of the total number of different species that live in that area. Both of these measures tell us something about the habitat type and quality, and often can be used as indicators of overall ecosystem health.

Environments that are richer in nutrients, water, and sunlight tend to have more productive plant communities, which in turn can support a greater abundance of animals. Therefore, the consistent presence of large numbers of animals in an ecosystem can be a strong clue that the ecosystem has plenty of raw materials needed for life to flourish. Just knowing the numbers is not enough, however. It is important to pay close attention to the long-term trends in numbers, and note whether populations seem to be increasing or decreasing. It also is important to look carefully at the makeup of the populations of animals you are counting. For instance, a population with a nice balance of old and young individuals would be very different from a population made up exclusively of old members. Nevertheless, an accurate estimate of animal abundance in an ecosystem can be a useful tool.

By itself, a count of animals won't provide enough information to gauge the health of an ecosystem. It also is very important to get an estimate of the variety of animals that are present. In general, when the ecosystem has a broader range of habitat and food types, it can support a wider diversity of animal species. Thus, a measure of diversity can be a very useful indicator of the quality of an ecosystem, with the presence of many different species often indicating a healthier ecosystem. The combination of estimating the number of animals and assessing the numbers and types of different species can be a powerful way to keep track of the environment.

The more we monitor species diversity and abundance in different ecosystems, the more easily we can be alerted to areas where populations of animals are declining steadily. A low diversity or low abundance does not necessarily mean a poor quality habitat. A system that is rebounding from a great disturbance, may be high in quality, but not yet have an established, well-developed community. However, declines of animals often are obvious symptoms of stress in an ecosystem. To make it easier to identify habitats and species that are in trouble, federal and state governments have created a set of categories for plant and animal species that have experienced recent population declines. *Endangered* species generally are those considered to be in imminent danger of extinction. *Threatened* species are those that are likely to become endangered within the foreseeable future. These categories are more than just names. Once a species becomes threatened or endangered, it is very aggressively protected under law. One of the most important aspects of protecting the organisms is the protection of the ecosystems that are critical to their existence.

Currently, the diversity of herps in the northeastern United States alone is high (see checklist on page 61). More than 21 species of frogs, 18 salamanders, 25 snakes, and 18 turtles inhabit the region. However, not all of the species are widespread throughout the entire region. Several of the herp species in the Northeast are extremely rare, and in many cases populations are shrinking rapidly. Some species are rare because they require specialized food, or rare habitats that only occur in a few places. Such conditions tend to exist in healthy ecosystems but are quickly lost in environments that have been disturbed. As increasing numbers of habitats become disturbed and destroyed, these specialized animals quickly are reduced to the few remaining refuges that will support them. Some species are rare in one part of their range, but more common elsewhere. For example, the tiger salamander and the eastern mud turtle occur in low numbers in New York, where they are listed by state officials as endangered and threatened respectively. However, New York appears to be at the northern end of the range of these two herp species. Farther south, these two species become much more common, and are not highly protected.

Many herps have become rare as a direct result of human activities. The five species of seaturtles that occur in eastern U.S. waters have been hunted, harvested, and unintentionally killed for so long that they all are threatened or endangered worldwide. In addition, the impact of hotels, houses, and other developments along turtle nesting beaches has contributed to their decline. In a similar fashion, hunting by humans has greatly reduced the number of rattlesnakes throughout their range over the past few hundred years. In northern regions, loss of preferred habitats seems to have constrained rattlers to inaccessible areas such as rocky ledges and swamps. Given the rapidly changing status of many different species of herps worldwide, and the simultaneous increase in human disturbance, it has become more important to increase our efforts to monitor and assess environmental health.

The most common first course of action is to monitor the species diversity within the area of concern. One way that scientists and volunteers measure diversity is by keeping a checklist of the species of plants or animals they observe. A checklist identifies all the species that are likely to be seen in a given area. The list generally is broken down into larger groups of animals, such as birds, amphibians, and mammals. Field biologists often create a customized checklist of the species in their region before beginning their studies. Using such a regional checklist can save time and effort, and also improve the accuracy of the data. If one of two similar species is not present in your region, it will not appear on the list, simplifying the process. Using a checklist also helps standardize data, making it easy for separate monitoring groups to compile and compare information.

Checklists also can be useful in learning about the behavior of animals. When used at different times at the same location, they can reveal information about timing of activity, breeding behavior, and migratory movements of organisms. Recording the presence of spotted salamanders or American toads at a pond in early spring is actually describing the location and timing of the breeding activity for these populations of amphibians.

Keeping a checklist not only provides important information about the diversity of animals in a region, but may also contribute to local conservation efforts. You may be able to team up with university or government researchers compiling an inventory of diversity within a particular area. Students can work with the researchers using the same checklist and the same procedures for making and compiling observations. By recording the number of individuals of a species, their location, the date, and other relevant information, you may be able to help scientists determine whether to be concerned about the status of the ecosystems, whether a species is common or rare, and whether local populations are healthy or disappearing.

Activity 8.1. Herp Checklist

Objective

Students will learn to use an animal checklist while making observations in the field.

Materials

- copies of "Herp Checklist for the Northeastern United States and Eastern Canada" (provided) (For other parts of the United States and Canada, contact wildlife departments at universities or state natural resources agencies to get checklists and species accounts.)
- copies of the four Herp Species Accounts (pages 133–145)
- field guides (see list below)
- plastic bags or folder to protect the checklist/species accounts (optional)
- clipboard

Procedure

Safety Tip: Have students read the section on safety (page 5) before going into the field.

1. Before going into the field, have the students use a field guide to modify the checklist to include the herps in your local area.
2. Students may want to enclose a copy of the checklist and relevant species accounts in a plastic folder to use when going into the field. Explain that checklists can be used each time they go out into the field and that naturalists often keep lifelong checklists recording the animals they have seen over their entire lives.
3. Have students record observations on the checklist during class and family field trips. They also may want to share the checklist for their region and their observations with local naturalists, schools, 4-H Clubs, and nature centers.

Field Guides

Regional

Harding, J. H. 1997. *Amphibians and Reptiles of the Great Lakes Region*. Ann Arbor, MI: University of Michigan Press.

Klemens, M. W. 1993. *Amphibians and Reptiles of Connecticut and Adjacent Regions.* Bulletin 112. Hartford: State Geological and Natural History Survey of Connecticut.

General

Behler, J. L., and F. W. King. 1979. *The Audubon Society Field Guide to North American Reptiles and Amphibians*. New York: Alfred A. Knopf.

Conant, R., and J. T. Collins. 1991. *A Field Guide to Reptiles and Amphibians: Eastern and Central North America.* 3rd. ed. Boston: Houghton Mifflin.

Herp Checklist for the Northeastern United States and Eastern Canada

Check	Taxonomic Name	Common Name
	Amphibians	
	FROGS	
☐	*Acris crepitans*	Northern cricket frog
☐	*Bufo americanus*	American toad
☐	*Bufo fowleri*	Fowler's toad
☐	*Hyla andersonii*	Pine barrens treefrog
☐	*Hyla chrysoscelis*	Cope's gray treefrog
☐	*Hyla cinerea*	Green treefrog
☐	*Hyla gratiosa*	Barking treefrog
☐	*Hyla versicolor*	Gray treefrog
☐	*Pseudacris brachyphona*	Mountain chorus frog
☐	*Pseudacris crucifer*	Spring peeper
☐	*Pseudacris feriarum*	Southeastern chorus frog
☐	*Pseudacris triseriata*	Western chorus frog
☐	*Rana catesbeiana*	American bullfrog
☐	*Rana clamitans*	Green frog
☐	*Rana palustris*	Pickerel frog
☐	*Rana pipiens*	Northern leopard frog
☐	*Rana septentrionalis*	Mink frog
☐	*Rana sphenocephala*	Southern leopard frog
☐	*Rana sylvatica*	Wood frog
☐	*Rana virgatipes*	Carpenter frog
☐	*Scaphiopus holbrookii*	Eastern spadefoot
	SALAMANDERS	
☐	*Ambystoma jeffersonianum*	Jefferson salamander
☐	*Ambystoma laterale*	Blue-spotted salamander

Ambystoma maculatum	Spotted salamander
Ambystoma opacum	Marbled salamander
Ambystoma tigrinum	Tiger salamander
Cryptobranchus alleganiensis	Hellbender
Desmognathus fuscus	Northern dusky salamander
Desmognathus ochrophaeus	Allegheny mountain dusky salamander
Eurycea bislineata	Northern two-lined salamander
Eurycea longicauda	Long-tailed salamander
Gyrinophilus porphyriticus	Spring salamander
Hemidactylium scutatum	Four-toed salamander
Necturus maculosus	Mudpuppy
Notophthalmus viridescens	Eastern newt
Plethodon cinereus	Eastern red-backed salamander
Plethodon glutinosus	Northern slimy salamander
Plethodon wehrlei	Wehrle's salamander
Pseudotriton ruber	Red salamander

Reptiles

SNAKES

Agkistrodon contortrix	Copperhead
Carphophis amoenus	Eastern worm snake
Cemophora coccinea	Scarletsnake
Clonophis kirtlandii	Kirtland's snake
Coluber constrictor	Eastern racer
Crotalus horridus	Timber rattlesnake
Diadophis punctatus	Ring-necked snake
Elaphe guttata	Cornsnake
Elaphe obsolete	Eastern ratsnake
Heterodon platirhinos	Eastern hog-nosed snake
Lampropeltis getula	Common kingsnake

Scientific name	Common name
Lampropeltis triangulum	Milksnake
Nerodia sipedon	Northern watersnake
Opheodrys aestivus	Rough greensnake
Opheodrys vernalis	Smooth greensnake
Pituophis melanoleucus	Pinesnake
Regina septemvittata	Queen snake
Sistrurus catenatus	Massasauga
Storeria dekayi	DeKay's brownsnake
Storeria occipitomaculata	Red-bellied snake
Thamnophis brachystoma	Short-headed gartersnake
Thamnophis sauritus	Eastern ribbonsnake
Thamnophis sirtalis	Common gartersnake
Virginia striatula	Rough earthsnake
Virginia valeriae	Smooth earthsnake

TURTLES

Scientific name	Common name
Apalone spinifera	Spiny softshell
Caretta caretta	Loggerhead seaturtle
Chelonia mydas	Green seaturtle
Chelydra serpentina	Snapping turtle
Chrysemys picta	Painted turtle
Clemmys guttata	Spotted turtle
Clemmys insculpta	Wood turtle
Clemmys muhlenbergii	Bog turtle
Dermochelys coriacea	Leatherback seaturtle
Emydoidea blandingii	Blanding's turtle
Eretmochelys imbricata	Hawksbill seaturtle
Graptemys geographica	Northern map turtle
Kinosternon subrubrum	Eastern mud turtle
Lepidochelys kempii	Kemp's ridley seaturtle

Malaclemys terrapin	Diamond-backed terrapin
Pseudemys rubiventris	Northern red-bellied cooter
Sternotherus odoratus	Stinkpot
Terrapene carolina	Eastern box turtle

LIZARDS

Eumeces anthracinus	Coal skink
Eumeces fasciatus	Common five-lined skink
Eumeces laticeps	Broad-headed skink
Podarcis sicula	Italian wall lizard
Sceloporus undulatus	Eastern fence lizard
Scincella lateralis	Little brown skink

Chapter 9

Herps and Their Homes

National Science Education Standards
Grades 5–8: Diversity and adaptations of organisms (Life Science)
Abilities necessary to do scientific inquiry (Science as Inquiry)
Grades 9–12: Abilities necessary to do scientific inquiry (Science as Inquiry)

There are many herp species across the United States, but you may have to look carefully to find them. They are distributed among a wide range of habitats: in dried leaves and soil in the forest, in grassy fields, in underground burrows, under rocks along stream edges, in ponds, and sometimes in trees. Even when looking for a single species, you may have to search several habitats at different times of the year. You will want to explore special nooks and crannies, or microhabitats, which are ideal for eggs staying moist in dry weather, or over-wintering.

Often several herp species will have different habits in a particular environment, allowing them to coexist rather than compete for limited resources. They each carve out their own ecological niche, or unique way of life. Sometimes the species do this by having different food items in their diets. Other species divide up a habitat based on their ability to tolerate dryness, climb vegetation, or burrow into the ground. Still others coexist by being active at different times of the day or year.

A large and diverse group of herps, such as the salamanders, may show a great deal of variation in their habitat requirements within a given region. For example, a distinct shift in the species of salamanders occurs as you move from the uplands down to the streams in the northeastern United States. Slimy and red-backed salamanders generally are found in dead leaves and under logs in the drier, upper parts of a forested hill slope, but their numbers decrease as you move down slope. They are replaced by dusky salamanders in the moister, cooler soils along the lower hill slope. Under rocks along the stream banks, the dusky salamanders are joined by two-lined salamanders, which move readily into the stream when alarmed. There you may find the totally aquatic mudpuppy and the very large hellbenders hunting for food along the stream bottom.

Many herp species become specialists on particular food items. This allows them to coexist with others in the same exact habitat. For example, tadpoles of different frog species may be so specialized that they feed at different depths and have mouthparts that are adapted for different foods, such as tiny invertebrates or algae on rocks. In the same way, adult frogs and salamanders may avoid competition by specializing on prey of different types and sizes.

Because of the broad diversity of herps and the variety of habitats they occupy, researchers seeking to catalog all of the species in a region have to develop a sampling plan before they begin. They need to establish standard methods for sampling and to think about all the habitats and microhabitats that need to be searched. They must consider how to look for nocturnal species as well as animals active during the day. They need to account for aquatic and terrestrial species as well as for animals that live underground. Finally, they need to account for different types of behavior at different times. Usually, to get a good idea of the species assemblage, or complete group of species, in the region, field biologists must conduct several different searches, at several times of day and during different seasons of the year.

One way to sample is to use a survey, that is, a record of what was seen on each outing. Good surveys include date, names of the observers, location searched, description of the habitats, and how long the search went on. Data recorded for the animals observed can be as simple as a count of species from a checklist. But most often a survey is distinct from a checklist because it includes the number of individuals of each species

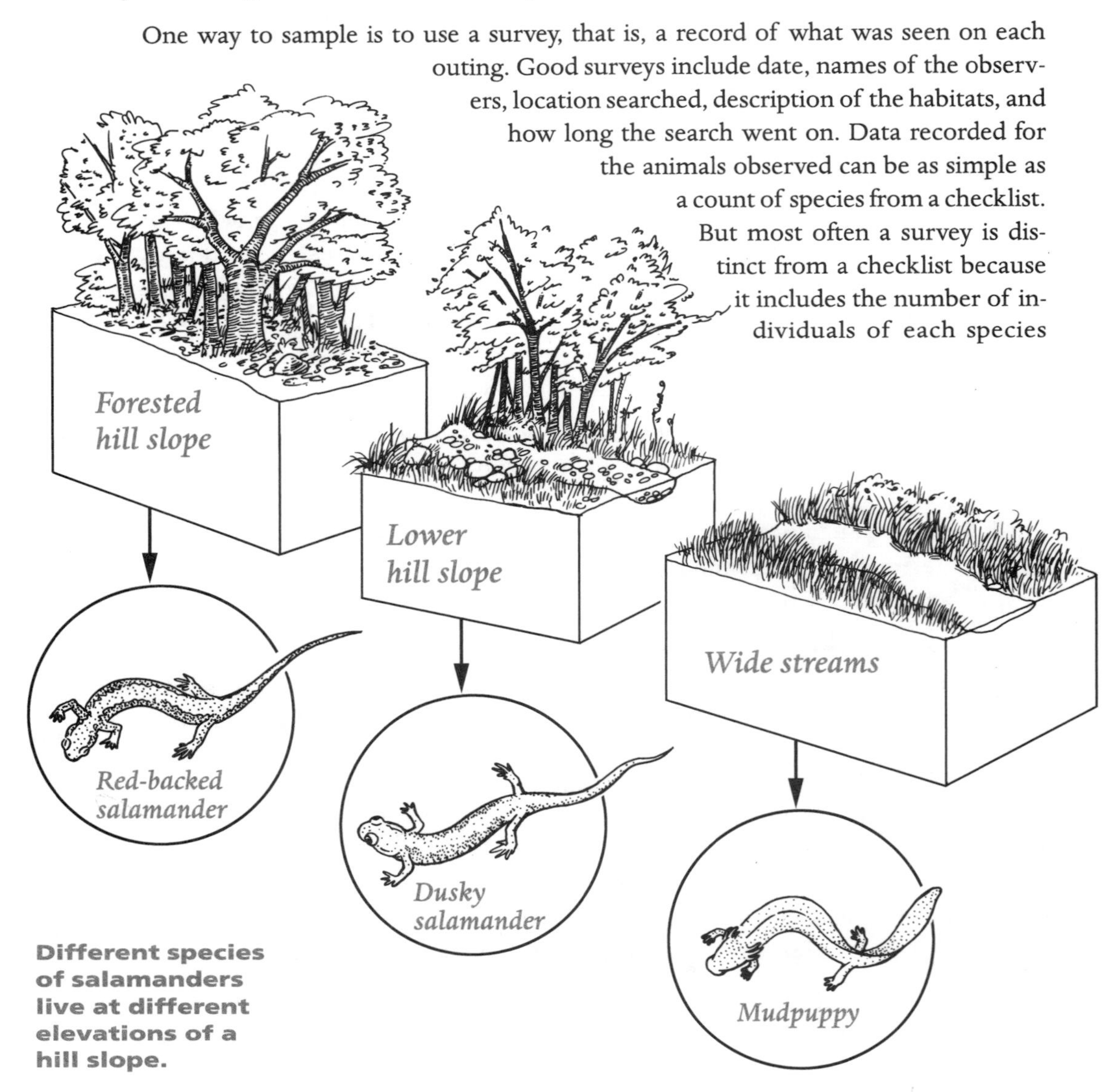

Different species of salamanders live at different elevations of a hill slope.

seen. In a visual survey, it is important to ensure that an individual is only counted once per survey.

A more complex survey might include records of what each individual was doing when it was observed (e.g., basking, resting, feeding, or mating). In any case, the types of data collected in a survey should be determined in advance and should reflect a well-planned study with specific goals. But always leave room for additional comments. You never know what fascinating tidbit you may uncover.

Many surveys are designed to conduct a census of populations or to provide estimates of the abundance of animals. For most herps, this is very difficult to accomplish by visual surveys. Because these animals often are well hidden, there may be very little relationship between observed and actual numbers. Therefore, many surveys, such as the frog-calling surveys, are designed to compare relative numbers from one survey to the next. Observers go to a pond where mating frogs are calling and try to estimate the number of individuals calling at that time. In most cases, only the adult males call, so at best such a survey should not be used to estimate the total number of frogs in the population, but instead to detect whether there are more or fewer frogs calling than in previous surveys. A survey conducted in subsequent years at the same time provides information about whether frog populations are declining, staying the same, or increasing.

Surveys, if done well, can provide a tremendous amount of information. From surveys, you can understand the relationships between herps and their habitats, the variation in their activities and behavior throughout the year, and even the relative abundance of animals from year to year.

Activity 9.1. Herp Walk

Objectives

Students will learn about the diversity of herps in an area, the variety of microhabitats herps occupy, and seasonal differences in activity.

Materials

- several large, flat boards that have not been chemically treated
- notebook and pencil
- flagging tape or other means to mark a trail
- field guides (page 60)
- copies of Herp Checklist (page 61)

Procedure

Safety Tip: Have students read the section on safety on page 5 before going into the field.

1. Have students mark a trail through several habitats: open fields, forest, streambeds, pond sites. The trail should encompass as many habitats as possible, including rocks along the stream edge, rotting tree stumps, and fallen logs. Set out some large, flat boards on the edges of the fields, near streams, and along the sides of the marked trail.
2. Have students go for walks along the trail with a checklist, species accounts, field guides, and a notebook set up to conduct a survey (date, names of observers, location, area searched, weather, length of search, and a description of the habitat).
3. Prior to beginning their herp walk, have students become familiar with "Care and Handling of Live Herps" (page 2).
4. Have students monitor the trail at regular intervals following the guidelines in the handout. Look under the rocks and logs, making sure to return them gently to their original positions. Check the boards during the day. First look on top for snakes that are basking in the sun, then look underneath to see what creatures may be enjoying this artificial cover. Return to the trails at dusk or night. Record the species found, the number of individuals observed, and their behavior.
5. Guide students in a discussion of the different species that they found at different times and in different habitats. Discuss what aspects of the species' behavior might influence where and when they are found. Students may want to share their observations with local scientists or conservation groups.
6. Be sure to pick up all boards at the end of the season, or when all surveys are completed.

Chapter 10

Food Webs

National Science Education Standards
Grades 5–8: Populations and ecosystems (Life Science)
Abilities necessary to do scientific inquiry (Science as Inquiry)
Grades 9–12: The interdependence of organisms (Life Science)
Abilities necessary to do scientific inquiry (Science as Inquiry)

Creatures in a natural community generally fall into categories that help define their roles in the ecosystem. Depending on what they eat or whether they are eaten, they may be predators or prey, and often are both. Herbivores, or animals that only eat plants, often are relegated to the role of prey. Carnivores, or meat eaters, are predators; omnivores eat both plants and living animals; and scavengers take advantage of decaying plants and animals. Many carnivores, omnivores, and scavengers, in turn, are preyed upon by other carnivores.

Although often unnoticed, herps play a major role as both prey and predators in aquatic and terrestrial ecosystems. In a freshwater pond, frogs and salamanders consume huge quantities of zooplankton, algae, and aquatic insects and are, themselves, eaten by large insects, turtles, fish, snakes, raccoons, and wading birds. On land, ratsnakes and black racers are fearsome predators of small rodents in fields and around farms, while gartersnakes and brownsnakes eat a lot of worms and insects. These snakes, in turn, are preyed upon by hawks, raccoons, foxes, and domestic animals, such as dogs and cats.

Researchers often develop diagrams or models to help them explain the relationships among creatures in a community. A common example is a food chain, which provides a model of the feeding relationships among the many creatures living in an area. In this model, plants are at the beginning and predators are at the end. Indeed, a food chain is a useful way to think about predator-prey relationships and the transfer of food energy in a community. Plants, including microscopic algae, capture light energy from the sun and convert it into living material, or biomass. These plants then provide food and habitat for many smaller invertebrates, which are in turn fed upon by amphibians, reptiles, and other predators. The concept of a food chain is a bit limited, however, in that it tends to simplify the rather complex associations among organisms. It also does not fully take into account the relationship of organisms to the flow of energy in an ecosystem.

Another way to think of these relationships is as a pyramid, where many small organisms at the bottom serve as food for fewer, and often larger, animals as you climb to the top. Studies of herps from a small South Carolina pond revealed amazingly that

Bullfrog Food Web

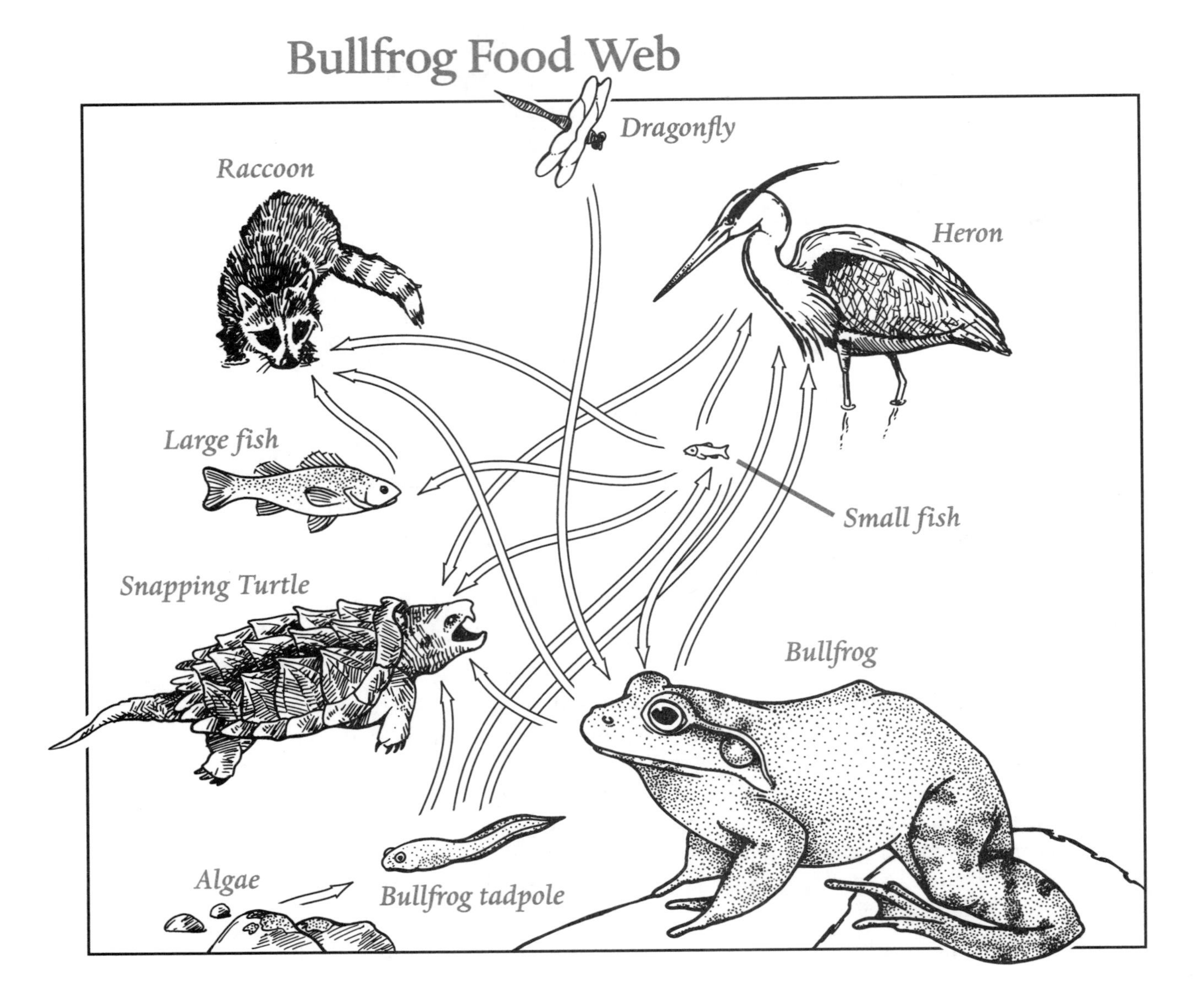

thousands of salamanders and frogs can live in the pond in a single year. These medium-sized animals are supported by huge quantities of tiny algae, plankton, and small insects, which form the base of the food pyramid. Higher up the pyramid in this ecosystem are perhaps hundreds of larger herps, fish, birds, and mammals that prey on the medium-sized salamanders and frogs. Theoretically, the energy in the pond is transferred up the pyramid, from the smaller to larger organisms, in a stepwise manner. The idea of a pyramid sometimes falls short, however, when you consider that some of the largest animals on earth, including elephants, giraffes, moose, and the Galapagos tortoise, eat only plants.

SCiLINKS
THE WORLD'S A CLICK AWAY
Topic: **food webs**
Go to: **www.scilinks.org**
Code: **HERP70**

In reality, the way in which food resources are distributed and consumed in an ecosystem is more complex than a food chain or a pyramid. It is better described as a food web, with multiple linkages among the different animals and plants. The feeding habits of painted turtles nicely illustrate why a food web is a useful depiction of predator-prey relationships in pond ecosystems. Early in life, a painted turtle will spend much of its time as a carnivore, preying on amphibian eggs, tadpoles, snails, and aquatic insects. As they grow, painted turtles shift their diets to include large amounts of vegetation, in some cases becoming predominantly herbivores. Painted turtles also are known to feed on carrion, or dead animals. Thus, a painted turtle fulfills various roles in the food web, including predator, herbivore, and scavenger. The true nature of the complex food web is further revealed when you consider that painted turtles can be an important part of the diet of predators, such as skunks, wading birds, raccoons, foxes, and other herps.

Activity 10.1. Discover a Pond Food Web

Objectives

Students will learn about the diversity of animal life in a pond and the interdependence among pond organisms.

Materials

- boots
- dip net with a fine mesh
- collecting jars
- white-bottomed collecting pan
- magnifying glass
- portable folding table
- copies of Herp Checklist (page 61)
- field guides (page 60)
- microscope (optional)
- *Pond and Stream Safari* by Karen Edelstein (optional; available through Cornell University, Media and Technology Services Resource Center, 7-8 Cornell Business & Technology Park, Cornell University, Ithaca, NY 14850. 607-255-2080. *Dist_Center@cce.cornell.edu; www.cce.cornell.edu/publications/catalog.html)*

Procedure

1. Have students fill the bottom of the collecting pan with water from the pond and place it in the shade to keep it cool. Dip the net into the pond and then carefully turn it inside out in the water in the collecting pan, releasing whatever has been caught. Take samples from various microhabitats, including the surface, the bottom, sediments, and within vegetation.
2. How many organisms can students pick out and identify? Even if they can't identify the species, can they guess the role the organisms might play in the pond

community? (For more information about sampling, identification, and natural history of pond insects and other invertebrates, obtain a copy of *Pond and Stream Safari* or a field guide to pond life.)

3. Have students take the samples of water and examine them under the microscope. How many different organisms can they count? How many plants? How many animals? Try collecting some pond scum. What kind of creatures do you find? Where do they fit in the food web?

4. Have students observe the rest of the animal life in and near the pond. Look for birds, fish, turtles, egg masses, and signs of animals, such as animal tracks, feces, underground burrows, and nests. Discuss what the different animals may be eating. From what they have seen, can the students surmise who are the predators and prey? Have them imagine being a specific reptile or amphibian living in the pond. What would they eat? How would they avoid being eaten?

5. Ask the students to draw a food web of the pond based on their observations.

Chapter 11

Upland-Aquatic Linkages

National Science Education Standards
Grades 5–8: Regulation and behavior (Life Science)
Abilities necessary to do scientific inquiry (Science as Inquiry)
Grades 9–12: The behavior of organisms (Life Science)
Abilities necessary to do scientific inquiry (Science as Inquiry)

Most animals have home ranges, or a minimum amount of habitat they require to meet their needs for food, protection, and reproduction. The size of a home range primarily depends on the species and the conditions and quality of the habitat. Animals may use parts of these territories only at certain times of the year—for example, when they are breeding, moving to obtain food, or moving to avoid freezing during winter months.

Many species of herps move from one habitat to another at different stages in their lives or at different times of the year. A very common type of movement is the dispersal of young animals into new habitats. In amphibians, this type of dispersal often is associated with a change in lifestyle, and even a drastic change in appearance. The gray tree frog is a good example of the many frogs and salamanders that disperse when young. It begins life as a tadpole in a pond. At this early stage, the frog is completely adapted for living in the water. It has gills for breathing, a long tail for swimming, and a specialized mouth for filtering and scraping algae. After approximately six weeks, legs sprout, the tail all but disappears, and its feeding and digestive systems change greatly. After undergoing this metamorphosis, the bright green juveniles are now capable of survival out of the water, and the froglets disperse into the adjacent upland area. There they find shelter, a new source of food, and a drastically different way of life. For the remainder of its life, the tree frog spends much of its time in woodlands, farm fields, and yards. Foraging on insects, spiders and snails, it can climb easily and often is found in shrubs or trees as high as 3 m (10 ft) off the ground.

Herps also travel between upland and aquatic sites for reasons other than dispersal of the young. Many amphibian and reptile species make regular treks to and from the water on a seasonal basis. Such periodic movements of animals from one portion of their home range to another generally are called migrations. The same gray treefrog that so completely reshapes its existence to escape the water early in life ultimately becomes a mature frog that must migrate back to the water each spring to mate. Such annual breeding migrations from the surrounding uplands to the ponds are extremely common among adult frogs and salamanders.

Aquatic turtles also move between ponds and surrounding uplands. However, their migration is almost the reverse of that of amphibians. Snapping and painted turtles, among others, live in ponds their entire lives feeding on frogs, fish, insects, plants, and other aquatic life. When it is time to reproduce, however, females leave the pond in search of appropriate sites to lay their eggs on land. They generally prefer open areas near the pond's edge but have been known to travel up to a kilometer (0.6 miles) to find a preferred nesting site. As soon as the small hatchlings emerge from their underground nests, they make their way toward the water where they begin their aquatic existence.

Perhaps the most remarkable herp migrations between land and water take place among the seaturtles. When it is time to breed, adult female seaturtles embark on what are sometimes tremendous journeys. They leave their feeding areas, which can be in northern waters, and may swim thousands of kilometers to a distant nesting beach in warm climates. There they crawl up on shore and lay their eggs in nests carefully dug in the sand. During their long lives, seaturtles are destined to repeat this process many times, emphasizing the importance of the linkage between land and water in the lives of most herps.

Activity 11.1. Sampling Migratory Herps with a Temporary Drift Fence

Objectives

Students will (a) observe herps as they migrate between aquatic and terrestrial habitats and (b) learn how to sample herps using a nondestructive, temporary drift fence.

Materials

- 2 10-m-long (33 ft) segments of landscape fabric or geotextile cloth (1 m [3 ft] in width)
- shovel or hoe
- 7 wooden stakes for each segment
- mallet
- staple gun
- data notebook and pencil

Procedure

1. Visit a small pond where eggs, tadpoles, or breeding adults are present and note the portion of the shoreline where frogs and salamanders may leave the pond, usually in a low-lying area that leads to other ponds or wet areas.

2. Lay out the cloth fences along the desired stretch of shoreline, approximately 3–5 meters (10–15 ft) from the pond's edge. Leave a space of about 2 m between the ends of the two fences. Select a site where large trees and shrubs won't interfere.
3. Leave 50 cm of cloth free at each end of the fence. Press the rest of the fabric into the ground so that the bottom edge is buried approximately 2 cm deep (1 in). If necessary, a shovel or hoe can be used to create a shallow groove. Go around any small tree stems.
4. Leaving the free ends alone, pound stakes about every 2 m (6 ft) to support the rest of the fence.
5. Use staple gun to attach the cloth upright to the stakes.
6. Take the free ends of the fence and bend them back, toward the pond, to form approximately a 45° angle with the rest of the fence. Attach the bent-back ends to additional stakes. The fence acts as a temporary barrier to herps that are moving out of the pond, and the bent-back corners will detain the animals a little longer to help you observe them.
7. Monitor the fence on a daily basis. If there are animals in the corners, gently assist them to the other side of the fence.
8. For each individual organism observed, record the date and time, the species, and the location of encounter, along with any desired comments.
9. Discuss with the students the data they have collected. Which species were most abundant? Discuss possible reasons: timing, feeding, breeding, habitat variations, etc.
10. It is important to remove the drift fence and fill in any holes at the end of the project to return the site to its original condition.

Reminder: Handle all herps carefully. Some herps are endangered or threatened and require permits for their collection.

Conservation and Management

Many scientists and conservationists are concerned that herps are declining in diversity and total numbers throughout the world, in some places at alarming rates. Many losses are not due to a single factor, but instead are the result of a host of impacts associated with human activities. The key to long-term protection for these animals is to understand and address each of the different threats. The five chapters in this section provide background on a variety of factors contributing to herp decline, including habitat loss and fragmentation, pollution, introduced predators and competitors, and inappropriate collection practices. The five activities are most appropriate for use with youth 12 years and older.

Chapter 12. Habitat Fragmentation

Chapter 13. Wetland Loss and Restoration

Chapter 14. Invisible Threats

Chapter 15. Species Conservation

Chapter 16. Crisis Intervention

Chapter 12

Habitat Fragmentation

National Science Education Standards
Grades 5–8: Populations, resources, and environments (Science in Personal and Social Perspectives)
Natural hazards (Science in Personal and Social Perspectives)
Grades 9–12: Natural resources (Science in Personal and Social Perspectives)
Environmental quality (Science in Personal and Social Perspectives)
Science and technology in local, national, and global challenges (Science in Personal and Social Perspectives)

Among the major impacts that human activities have on wildlife is the broad-scale fragmentation of their landscape. Once large, continuous stretches of forests, grasslands, and swamps existed uninterrupted by cities. Roads, railroads, and power line right-of-ways now crisscross the landscape. Today vegetated and aquatic habitats often are found in small, isolated patches separated by agricultural fields, paved surfaces, and buildings. Such fragmentation has a tremendous impact on wildlife that were once dependent on those connected habitats.

Many frogs, salamanders, snakes, lizards, and turtles are small, ground-dwelling creatures that crawl across the land and often migrate from one habitat to another. As such, they are particularly sensitive to the impact of landscape fragmentation. When two formerly connected habitats become separated by a clearing or new development, the herps that need to traverse these human-made gaps are exposed to new obstacles and dangers.

A common example is a road that separates a frog or salamander breeding pond from adjacent woods where the adult animals live. The first few rainy days and nights of spring are the cue for thousands of adult frogs and salamanders to leave their upland wintering burrows and migrate into nearby ponds to mate. Often, however, a road interrupts their pathway to the pond. Before the roads were present, these adults might have successfully reached their breeding habitat and produced millions of eggs. Instead, hundreds or thousands of adults may be run over by cars while attempting to cross these roads.

Turtles face a similar threat when they are forced to cross roads while foraging, seeking a mate, or migrating to a nesting site. In addition to cars, a 20-cm-high curb can be a monumental barrier to a turtle, forcing it to detour for long distances before finding a place to scale the obstacle. Even if herps that enter roadways are not run over, they may become trapped or suffer from long-term exposure to drought or heat.

Fields and lawns also may be inhospitable to herps on the move. They have less organic matter to hold moisture and generally have less vegetation to provide shade and cover. As a result, herps on lawns or fields may be more exposed and more vulnerable to predators and to overheating.

The separation of habitats by inhospitable terrain also may have serious long-term effects on the survival of a species within a region. During the course of a species' existence, individual populations may fluctuate between periods of great abundance and periods of greatly reduced numbers. Sometimes environmental hardships, such as flooding or droughts, can wipe out entire communities of herps that were previously flourishing. One way a species can persist during such hardships is to relocate to a less stressful, nearby environment. When the stress is later reduced, the original habitat can be re-colonized through migration. However when habitats become separated from each other by inhospitable habitat, migration and re-colonization may become impossible. This lack of ability of animals to move readily among habitat patches is thought to be a leading contributor to the widespread decline of many herps.

Many conservation and management techniques can lessen the problems of habitat fragmentation. Protection, not only of the primary herp habitats, but also of movement corridors between habitats, is critical. Vegetated, undeveloped "buffer areas" around ponds can be established to protect the habitat needed by frogs, salamanders, and turtles. Migration pathways between wetlands and uplands and among adjoining habitats should be protected through the development of greenways, or vegetated corridors. Where interfering roads are already in place, alternative methods will be needed. With cooperation from the town and highway department, it may be possible to remove curbs or construct culvert-type tunnels under large highways to allow migrating animals to pass beneath the road. At a minimum, signs posted along the roads where herps move can educate humans to be aware of their fellow travelers and encourage them to drive more carefully.

Activity 12.1. Herp Crossing Alert!

Objective

Students will become engaged in a real-life conservation activity that helps herps cross the gaps between fragmented habitats.

Materials

- waterproof paint and brushes
- stencils
- plywood and wooden posts
- saws, hammers, nails

Procedure

1. Locate an amphibian breeding pond or marsh that has a road running alongside that interferes with migrations. Talk with the highway department about the problem of roads and spring migrations of amphibians. Inform them of your intentions and get their permission to post signs along the road near the pond or marsh.

2. Have students make two herp crossing signs, one for each direction of travel along the roadway. Cut the signs out of plywood. Write a polite warning, such as "Frog Crossing—Please Drive Slowly!," using waterproof paint in strongly contrasting colors that will show up well in the headlights of a car. Remember that most breeding migrations take place on rainy nights.

3. Visit this site on a rainy night in early spring when you begin to hear the frogs chorusing. Identify the stretch of road where most amphibians are crossing to enter their breeding site. Place the signs along both sides of this stretch of road.

4. Create a simple, one-page flyer that explains the importance of amphibian migrations and why neighbors need to drive carefully in certain areas to avoid squashing animals. Distribute these to residents who live near the breeding pond and are most likely to be using this section of road.

Chapter 13

Wetland Loss and Restoration

National Science Education Standards
Grades 5–8: Populations and ecosystems (Life Science)
Populations, resources, and environments (Science in Personal and Social Perspectives)
Natural hazards (Science in Personal and Social Perspectives)
Grades 9–12: Environmental quality (Science in Personal and Social Perspectives)
Natural and human-induced hazards (Science in Personal and Social Perspectives)

Many species of amphibians and reptiles require wetlands for at least part of their life cycles. A wetland is a special habitat where the water table is at or near the soil surface, or the land is covered by shallow water for parts of the year. There is a wide variety of wetland types, ranging in size and permanence from temporary vernal pools, to forested swamps, to the marshy shorelines of large lakes and rivers. All of these types of wetland habitats support thriving amphibian and reptile communities.

Plants within the wetlands are major features that contribute greatly to the vitality of amphibian and reptile populations. From the tiniest algae to the towering stands of cattails, wetland plants provide food for tadpoles, aquatic insects, and zooplankton, which in turn provide food for larger herps of all kinds. Leaves and stalks provide cooling shade, cover from terrestrial and flying predators, and resting places for calling frogs. Underwater the stems also provide attachment sites for eggs and act like a dense forest in which juveniles and adults hide from predatory fish, wading birds, and other herps. In general, the wetland plants form the basis of some very complex and productive communities upon which many populations of herps depend.

In the United States, over the past 200 years, more than 50 percent of the original 220 million acres of wetlands have been destroyed for agriculture and land development. These lands have been drained, cleared of vegetation, and filled with soil. The tremendous loss of wetlands has resulted in an unfortunate loss of prime habitat, and undoubtedly led to great reductions in populations of herps and other wildlife. The tremendous loss of wetland habitats has prompted great concern and has resulted in changing our laws. Starting in the 1970s, federal, state, and local governments enacted regulations to protect our remaining wetlands from further development and destruction. These laws have been successful in slowing the rate of wetland loss. However, wetlands are still under pressure from ever-increasing human activities. According to the U.S. Fish and Wildlife Service, an average of 117,000 acres of wetlands each year were lost to development between 1985 and 1995. This trend continues today.

A major problem contributing to wetland loss is the lack of understanding of how wetlands function and a lack of recognition of their role in supporting wildlife. When there are multiple wetlands in an area, humans tend to view some as surplus. The notion that, if there are two wetlands nearby, only one is necessary for wildlife comes from the practice of looking at each site individually. However, each successive development has cumulative effects across the larger landscape. For many herps, each wetland loss represents an immediate reduction in the amount of available breeding habitat and food resources. Also, the apparently "extra" wetlands often are critical in providing overflow habitat during times when populations are growing or providing refuge during droughts and other unfavorable environmental conditions. Without the availability of these nearby habitats, local herp populations may decline over time, and may even be driven to extinction. Thus, when a small wetland is drained and filled to allow the construction of, say, a shopping center, the impact on wildlife may reach far beyond the immediate realm of the single wetland.

Recently there has been an attempt to reverse the historical decline of wetland ecosystems. Efforts are underway to restore wetlands that have been cleared or drained and to create new wetlands. Such efforts are having mixed success. In particular, scientists are finding it difficult to recreate the original hydrology, or water regime, of the wetlands, and to reestablish the wetland plant communities. The science of restoration is still very new; thus, it is important to protect the wetland resources that we have.

A worthy goal is preserving wetlands on a larger landscape scale. Rather than preserving individual sites, groups or clusters of wetlands in an area can be preserved or created to ensure the long-term survival of wildlife populations in the area. In a pilot study in South Carolina, four small ponds were created to provide an alternative habitat when a neighboring pond and its wetland were drained and filled. Monitoring over the next several years indicated that many of the former pond's frog and salamander residents eventually found their way to the new refuge ponds and were breeding successfully.

Activity 13.1. Frog-Friendly Ponds

This is an intensive and lengthy restoration project, best conducted during the summer and fall.

Objective

Students will improve the quality of a pond habitat for amphibians.

Materials

- digging tools (shovels, pickax)
- hay bales

- 1 m (3 ft) long wooden stakes
- wetland plants
- wetland sediment

Procedure

Safety Tip: Consider having youth wear life jackets when working around the edge of deep ponds. Younger children should be supervised by an adult.

1. You can improve the suitability of a pond or lakeshore for herps by creating a healthy wetland plant community along its shoreline. Find a willing farmer or a nature center, preserve, or golf course manager who is interested in enhancing his or her wetland or pond for herps and other wildlife. Most wetland restorations are regulated by federal and state governments. Depending on the scale and nature of your project, you may need to check with your state natural resource agency regarding permits or other regulatory considerations.

2. Many farm ponds have very steep shorelines that inhibit wetland plant growth. The shoreline needs to be graded so that the bank is not too steep and the water is sufficiently shallow for wetland plants to survive. You can grade the bank by hand, using shovels and pickaxes to create a gentle angle of 25° or less.

3. While grading, it is important not to let sediment cloud up the pond or move into downstream water flow. This is best accomplished by grading during low-water periods in late summer. For extra protection, use wooden stakes to anchor hay bales at the toe of the slope to trap any loose soil.

4. Next, select plant species that can tolerate the average water depth and flooding regime they will experience throughout the growing season. A variety of emergent wetland plants, including pickerelweed, rushes, and cattails, grow well in wet soils or flooded up to about 30 cm (1 ft). Water lilies do well in more continuously flooded habitats and in water up to 2 m (6 ft) deep. Ask your local cooperative extension office or nature center about which plant species are most suitable for your climate, soil types, and water chemistry. Be careful not to choose invasive plants, such as purple loosestrife or common reed grass, which may take over the wetland, crowd out species that are more beneficial to wildlife, and then spread to other wetlands. Plants can be purchased from a nursery or transplanted with permission of the landowner from nearby wetlands. Planting is best done in early spring or fall when the plants are dormant so that the shock of transplanting will be minimized.

5. An alternative to transplanting is to start a wetland using fresh surface sediments obtained from a nearby wetland site. This sediment will contain thousands of seeds that will germinate in their new environment. It also contains rhizomes and roots that may grow into new plants. Be sure to get permission from the

landowner first before collecting several buckets of wetland soil. In early spring, before germination has begun, spread a shallow layer of sediment, approximately 5–10 cm (2–4 in) deep, across the newly graded shoreline. A thin layer of straw spread over this surface will help to reduce erosion and loss of the seeds should there be a strong rain.

6. Whether you use plants or sediment, the shoreline wetland will need maintenance until the plants become established. Water the plants and sediment frequently until the growing plants develop a healthy root system. Once the new wetland plants become established, remove the hay bales.

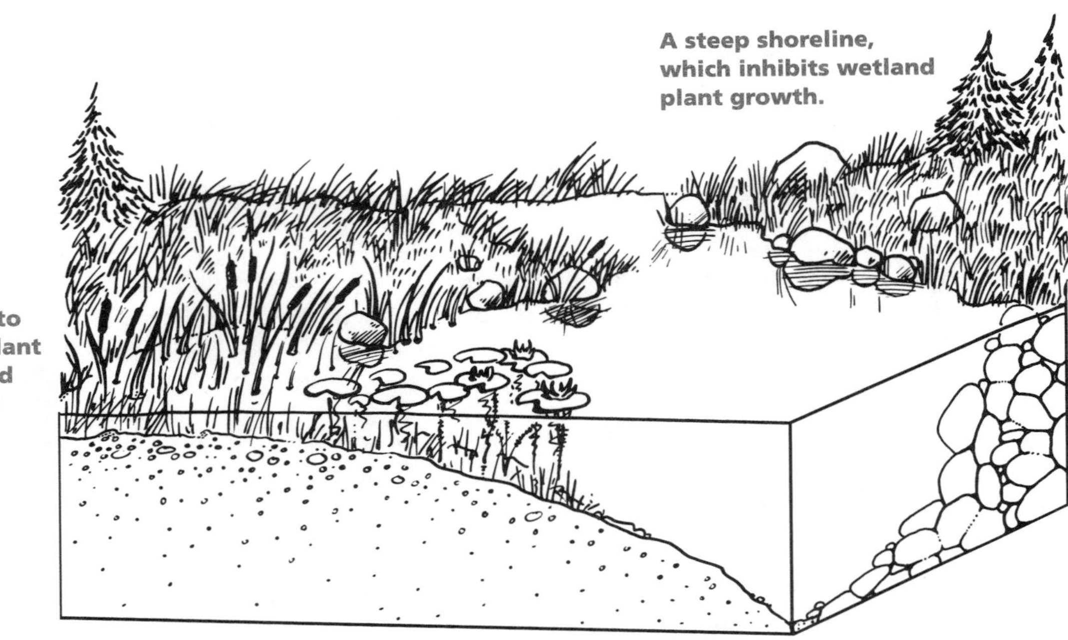

Chapter 14

Invisible Threats

National Science Education Standards
Grades 5–8: Populations, resources, and environments (Science in Personal and Social Perspectives)
Natural hazards (Science in Personal and Social Perspectives)
Grades 9–12: Environmental quality (Science in Personal and Social Perspectives)
Natural and human-induced hazards (Science in Personal and Social Perspectives)

It is easy to see how the destruction of a pond or the paving of a forest can directly affect herp populations. However, human activities frequently impact herps in many serious ways that are less visible and obvious than outright habitat destruction. These threats include direct and indirect changes to the physical environment, as well as to the biological communities that are important to herps. Threats also can express themselves at different scales, from very sudden local impacts to long-term and far-reaching negative impacts around the world.

For many species of herps that depend upon wetlands and aquatic sites during their lives, an important condition of their existence is the hydrologic regime of the site. The hydrologic regime, which refers to the timing and frequency of water fluctuations, is influenced by rainfall and snowmelt and generally is affected by the physical characteristics of an aquatic site. Water moves into and out of the site overland in creeks and rivers and underground through the transport of groundwater. Under natural conditions, water levels in streams, rivers, and lakes generally fluctuate at different times throughout the year. Many animals and plants are highly adapted to these natural cycles and, in many cases, require water movement or predictable fluctuation to meet their needs.

However, the hydrologic regime of many streams, rivers, and lakes has been altered by dams and levees or through excessive water withdrawals for irrigation or public water supply. In some cases, water is extracted from rivers and transported to cities hundreds of kilometers away, completely altering the river ecosystem and its associated wetlands. Excessive withdrawal of water from aquifers has similarly affected the hydrologic regime in shallow ponds of eastern New York, Colorado, and elsewhere. An interesting phenomenon occurs in some resort areas where stream water is withdrawn during winter for artificial snow production. Such manipulation of water levels can have devastating affects on herp communities. The alligators in the Everglades experienced extreme declines in nesting success in the early 1980s as a result of intentional changes in water levels to help control the flow of water. Alligators, nesting in the same manner they had for millions of years, were unwittingly placing their eggs in areas that were soon to be flooded. In this case, the declines appeared somewhat mysterious until the cause was discovered to be related to unnaturally high water levels.

Another major threat to aquatic habitats and their herp communities is the unobserved input of chemical pollutants into the water and sediment. Chemical pollution in the form of acid rain and pesticides affects the health and reproduction of herps, in some cases contributing to disappearances of entire populations. Although many pollutants are initially deposited on the ground or released into the atmosphere, they often are carried eventually into streams and lakes. When rain falls on the piles of mining wastes outside of mines, acids leach out along with large quantities of aluminum, zinc, and other heavy metals, and wash downhill into aquatic environments. In high concentrations, heavy metals are extremely toxic to most animals.

Another threat to herp populations that has received more recent recognition is the introduction of nonnative species, sometimes referred to as "biological pollution." In many instances, organisms are deliberately introduced to serve a human need. For example, in farm ponds, streams, and lakes fish stocking is a common practice that serves as a source of food and recreation. Most often, the game fish are high-level predators, which can be devastating to the native herps. In some habitats introduced species sometimes become more of a problem than chemical pollution, because they are able to reproduce on their own and can be very difficult to control. Even other herps that are transplanted into new environments can be disruptive to the native members of the community both as predators and competitors. The American bullfrog is native to the eastern United States, but has been introduced throughout western North America, Puerto Rico, Italy, and elsewhere. In many of its new habitats, it is proving to be incredibly prolific and voracious, and is outcompeting many local frog species.

Indirect threats of human presence also can be harmful to local herp communities. Many predators of herps prosper due to their association with humans. Raccoons, skunks, and opossums reach very high densities in urban and suburban landscapes, in part because they supplement their diet by dining in trash cans. These animals also readily eat amphibians and reptiles that live within their territories. Similarly, domestic cats and dogs are fierce predators on many ground-dwelling organisms. Outdoor cats commonly prey upon and devastate populations of frogs as well as mice, shrews, and other small animals living as far as 2 km (1.2 mi) from the pet's home. Because most pets are well fed, they can persist in stalking some prey species who normally deter hungrier, natural predators, such as foxes or hawks, by waiting them out. Dogs also will dig up amphibian burrows and turtle nests. Because humans live just about everywhere, there are few areas left where domestic pets and tagalong predators are not a threat.

Activity 14.1. Threat Assessment

Objective

The students will conduct a threat inventory to assist in the management plans of a nature center or preserve.

Materials

- 1:24,000 scale U.S. Geological Survey topographic maps showing the nature center or preserve and its surroundings (available from camping stores and on the Internet at the U.S. Geological Survey website: *mac.usgs.gov/mac/findmaps.html*)
- property tax maps of the area surrounding the preserve to show locations of homes and businesses (available at town halls)

Procedure

1. Find a nature center or wildlife preserve that is willing to work with the students in developing a management plan for herps.
2. Have students identify the boundaries of the preserve on the topographic map. (Preserve staff may be able to help.) Highlight the location of wetlands and ponds within the preserve that may be good habitat for herps. Also highlight the ponds and networks of streams that traverse the preserve, including where they continue upstream or downstream off of the preserve property.
3. Use the topographic and property tax maps as well as other available information to locate sources of threats to the preserve's herp populations. Include location of the following:

- Upstream dams or water removal operations on the streams that may be altering the hydrologic regime.
- Nearby municipal wells that could potentially influence groundwater levels and hydrologic regime of the ponds or streams.
- Residential homes within 1–2 km (1 mi) of the ponds that might have pets that could act as predators. (It might be possible to drive around the adjacent roads and record homes with doghouses.)
- Sewage treatment plants or other industrial waste sources, upstream from the preserve, that may be a potential source of pollution.

4. Have the students summarize their findings and discuss them with the preserve manager. See if they can come up with ways to use this information to protect the preserve. For example, they may want to design a brochure for preserve visitors or create a public service announcement for local residents.

Chapter 15

Species Conservation

National Science Education Standards
Grades 5–8: Populations, resources, and environments (Science in Personal and Social Perspectives)
Natural hazards (Science in Personal and Social Perspectives)
Grades 9–12: Environmental quality (Science in Personal and Social Perspectives)
Natural and human-induced hazards (Science in Personal and Social Perspectives)

Collection of herps from the wild may go unnoticed by the public. However, such practices deserve serious consideration, mainly because they are occurring at a high rate and can have serious effects on herp populations. There are some advantages to collecting animals from the wild. On the positive side, herps in zoos, aquaria, and nature centers contribute to education, increasing the public's awareness and appreciation for wild creatures. A carefully planned display of exotic and local species can have an incredible impact as an education tool and as a means of conveying important conservation information. In addition, by raising animals in controlled environments scientists can learn about animal behavior and other biological information that is critical to maintaining healthy populations in the wild.

Recently, more emphasis is being placed on propagating animals in captivity. Zoos are playing a major role in captive breeding programs, which provide stock for other zoos and also maintain genetic material for species that have become extremely rare or even extinct in the wild. Such breeding programs also may relieve the pressure on wild animals that are collected as a source of chemicals that have medicinal benefits for humans. For example, secretions from the mucous on the skin of some amphibians inhibit fungal and bacterial growth. Herpetologists are developing methods for breeding the frogs and "milking" the secretions from them without harming them. The biochemicals that come from the frogs also provide invaluable models for scientists who are developing new medicines, which further relieve the pressure to collect wild animals.

The uncontrolled removal of herps from the wild can have severe consequences for the animals and their environment. Harvesting eggs or adult herps for food or other commercial products is a worldwide problem that has driven many species close to extinction. The American alligator once was brought to the brink as a result of the fad for alligator bags, belts, and boots. It was only through careful restrictions in recent years on the use and sale of alligator skins that these reptiles were saved.

Local herp populations also suffer when humans view these animals as a threat. Snakes always come to the front of the line when it comes to animals that are feared and often despised. In many regions, people have worked actively to remove snakes from their surroundings. "Rattlesnake roundups," gatherings to eradicate local rattlesnake populations, have long been a tradition in parts of the southwestern United States. Indeed, there is justification for fearing some snakes. Unfortunately, many people are unaware of the benefits provided by most species of snakes, and indiscriminately kill any snake that they encounter. As a group, the snakes have been severely affected by being singled out.

Similarly, the collection of herps for the pet trade has had a serious impact on several species. Many collectors have a poor understanding of the factors maintaining wild populations, and persist in collecting until they have eradicated a local population. In addition, because many exotic herps bring high prices as pets, some collectors knowingly continue to deplete wild populations. The eastern indigo snake of the southeastern United States is one such victim that was overcollected, mainly because it made such a gentle and attractive pet. After decades of stringent regulations and intense conservation efforts, the eastern indigo snake still is considered a threatened species.

Another problem associated with the pet trade is high mortality of animals due to pet owner ignorance and poor care. Lizards and turtles frequently are bought as pets for children because they are thought to require less attention than dogs or cats. The pets often are handled excessively, underfed, or exposed to extreme temperatures, and die. When a family loses interest or can no longer support a pet herp, many are released back into the "wild." The environmental staff in New York City's Central Park report that they deal with hundreds of such releases every year. Herps released in the more northerly states sometimes are unable to survive the cold and die from exposure. Farther south, the released herps may have a better chance of surviving the weather, but they usually are not where they belong. The high mortality rate of captive and released pets ultimately helps stimulate the demand for more animals from the wild.

Adequate conservation of herps is going to require considerable effort using different approaches. Habitats where herps live need to be protected. Techniques for proper care and handling of pets need to be developed, including the potential for recycling pets to new owners. Furthermore, regulations to reduce the illegal, uncontrolled sale of herps need to be enforced. But above all, education is needed to increase awareness about the benefits of herps in their natural habitats.

Activity 15.1. Recycling Unwanted Herp Pets

Objective

Students will engage in conservation and outreach activities to help improve the herp pet trade and pet care.

Procedure

1. Work with your local Society for the Prevention of Cruelty to Animals (SPCA), animal shelters, and pet stores to build a program for the "recycling" of unwanted,

exotic herp pets. Help the SPCA set up appropriate aquaria to house the returned animals.

2. Create a poster campaign for your local schools to educate both children and their parents about the importance of good herp pet care and what to do with their unwanted pets.

Chapter 16

Crisis Intervention

National Science Education Standards
Grades 5–8: Populations, resources, and environments (Science in Personal and Social Perspectives)
Natural Hazards (Science in Personal and Social Perspectives)
Grades 9–12: Environmental quality (Science in Personal and Social Perspectives)
Natural and human-induced hazards (Science in Personal and Social Perspectives)

When the size of a population drops below a critical level, the remaining individuals may not be able to produce enough offspring to reverse the decline. Sometimes even small pressures on such vulnerable populations can push them over the edge. When this happens to an entire species, it faces the possibility of extinction.

Conservationists use a combination of legislation and direct intervention strategies to counteract species declines. In 1973, Congress passed the Endangered Species Act to protect species in danger of extinction. The act is also proactive in that it protects threatened species—those plants and animals that may soon become endangered. Many states have added another layer of protection by developing their own lists of animals and associated regulations. Intervention techniques are designed to protect animals and decrease mortality at many different stages throughout their lives.

For turtles, nesting is a critical phase that warrants special attention. In some situations, turtle nests are protected as part of a broader habitat preservation program. For the gopher tortoise of the southeastern United States, the conservation of sandy ridge and flatwoods habitats also helps protect the turtles' nesting habitats. In other situations, it may be necessary to protect individual nests. Seaturtle nests along tropical beaches suffer from heavy predation by wild and domestic animals and by humans, who use the eggs as a source of food. Some seaturtle conservation efforts are directed at locating and marking all the nests along a beach during the periods of peak egg laying. Park guards and volunteers then patrol the beaches to ensure the safety of the nests while the eggs incubate underground. Fences may be constructed around each nest to keep out persistent predators. Some seaturtle conservation programs physically transplant the freshly laid eggs to a protected hatchery, where the eggs are artificially incubated. When the eggs hatch, the newly emerged young are transported directly to the water's edge.

Hatchlings also are a highly vulnerable stage in the life cycle of most herps, and many intervention strategies work to reduce their mortality. "Head start" programs not only artificially incubate the eggs but also feed and maintain the hatchlings until they have grown to a large enough size to increase their chances for survival. Such head start programs appear to be successful in the reestablishment of the threatened Blanding's turtle in eastern New York and the Plymouth red-bellied turtle in southeastern Massachusetts. But there is some concern that such artificially isolated and reared animals may not learn the critical skills and information they need for long-term survival.

Many herps remain juveniles, or nonreproducing members of the population, for years. Human intervention and assistance can increase survivorship during this period. During this extended period of rapid growth, young turtles face many hardships, and mortality can be high. A particular example of human intervention having successfully decreased mortality of young seaturtles is in the case of cold-stunned seaturtles in the northeastern United States. Each summer, juvenile seaturtles move into shallow coastal waters from Virginia to the Gulf of Maine. These turtles are predominantly young loggerhead, Kemp's ridley, and green seaturtles, which spend the warmer months feeding on bottom-dwelling animals and plants. In October, when water temperatures drop, the turtles leave the cold waters and migrate southward to their over-wintering areas. Each year, however, anywhere from a few to a few hundred young turtles make a mistake and remain in the northern waters too long. In the cold water, the turtles become sluggish and can no longer swim. From November through December, these cold-stunned animals wash up along the shorelines of New York and New England, many lying dazed on the sand, destined to die from exposure.

Cold-stunning is a natural occurrence in the life of some sea turtles. However, for populations already severely depleted, the loss of several hundred individuals can represent a powerful blow to the entire species. For this reason, rescue groups have been successfully rehabilitating hundreds of cold-stunned seaturtles along much of the northeastern coast. The revived turtles usually are kept over winter at the nearest aquarium or transported south to warmer waters. The rescue groups are a successful example of the collaboration among citizen volunteers, who walk beaches daily looking for stranded turtles; nonprofit organizations, which coordinate the volunteers' efforts and provide a central receiving station; and government agencies, which provide technical access and critically needed funds.

Adult herps also may need intervention to reduce mortality in order to preserve their species. Over-harvesting is a major cause of species declines. If a herp population is at a low but sustainable level, some controlled hunting may be permitted. For example, after vigilant protection for several decades, the alligator populations of the southeastern United States have returned to sufficiently healthy levels to permit some restricted hunting. However, controls on harvesting are not always easy. Hunting to deliberately exterminate local venomous snakes is a common practice, and protection efforts require a combination of legislation, education, and enforcement to be successful.

Seaturtles are at such globally low numbers that all harvesting has been made illegal. Efforts even are being made to protect these extremely rare creatures from accidental sources of mortality. Shrimp trawling and other fishing activities abound in coastal waters where adult seaturtles are found. For many years, seaturtles have been caught and drowned in the nets used to catch shrimp, flounder, and other fish. The use of turtle excluder devices (TEDs) is helping to reduce the drowning of turtles. These trapdoor devices permit the turtles and big fish to swim out of the nets without releasing the intended catch. The use of TEDs has been successful in decreasing fishing-related mortality in seaturtle populations. However, other accidental deaths still occur frequently because activities of both seaturtles and humans tend to overlap in nearshore environments. Turtles coming to the surface to breathe are struck all too often by high-speed boats that crowd the waters during summer months. These collisions frequently prove fatal to the turtles.

Preventing species from going extinct takes considerable effort and persistence. Numerous organizations and citizen groups are using a diversity of intervention strategies to protect or restore herps in danger of extinction. Obviously such efforts require funding and resources. But the real key to success is the numerous volunteers and professionals who donate their time and efforts to help avert crises.

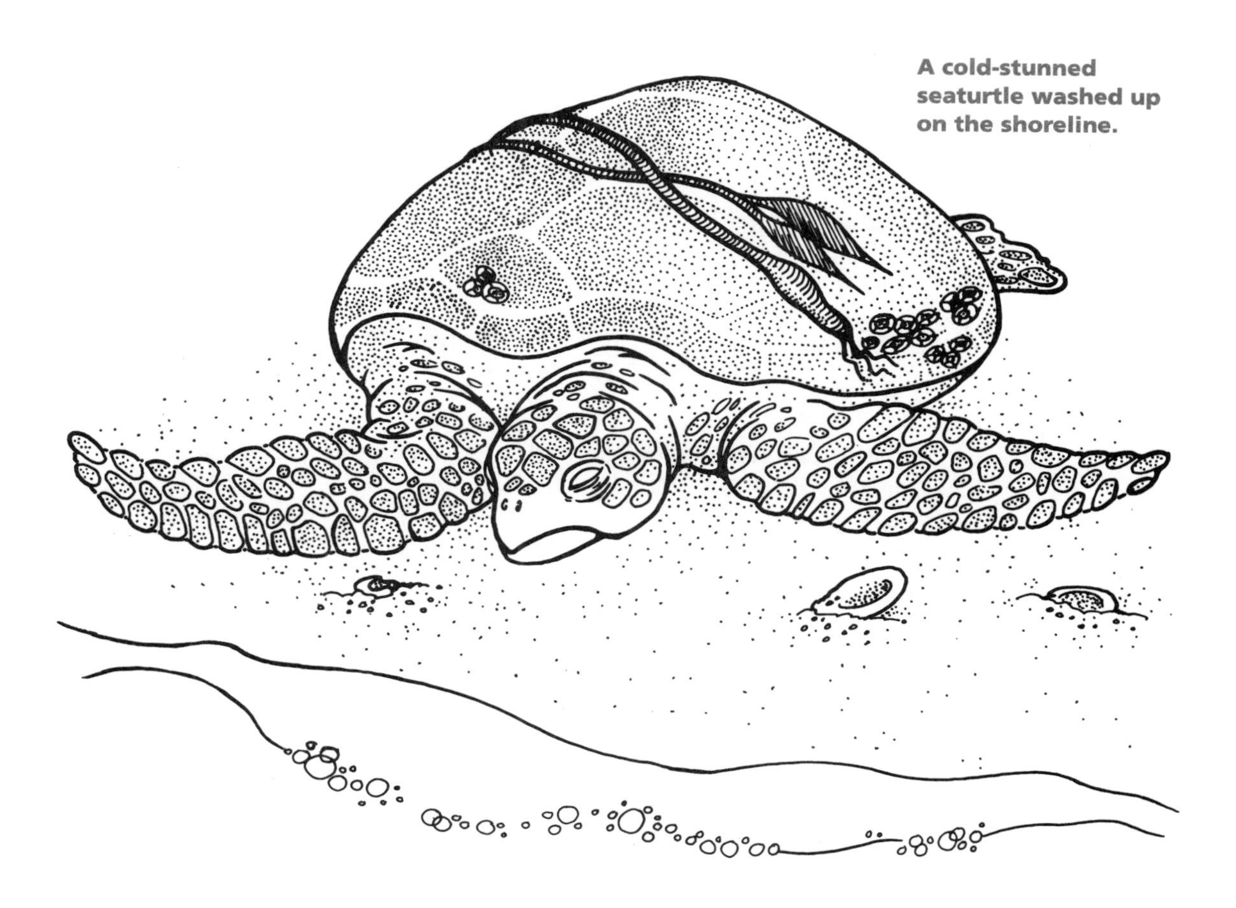

A cold-stunned seaturtle washed up on the shoreline.

Activity 16.1. Seaturtle Head Start and Rescue Programs

Objective

Students will learn about local intervention efforts to increase herp survivorship. If appropriate, the students can participate in these programs as volunteers working with the program leaders.

Procedure

1. Find a turtle head start or seaturtle stranding program in your area. Several programs located in the northeastern United States are listed below. Alternatively, search the Web for programs in your area by using key words of "turtle head start" and "turtle stranding." If you are not near a coastal area with seaturtles, you might contact your state natural resources agency or a local university to ask about other herp conservation efforts in which you might participate.

2. Contact the program coordinator or use the program's website to get information about the program's goals, activities, evidence of success, and challenges. If possible, compare information from several different intervention programs. Discuss with the students the important role that volunteers play in making these herp intervention programs succeed.

3. If appropriate, have the students participate as volunteers in one or more of the activities coordinated by a local herp intervention program, such as the following:

   The Riverhead Foundation for Marine Research and Preservation
   428 East Main Street
   Riverhead, NY 11901
   631-369-9840
   www.riverheadfoundation.org

   Plymouth Red-Bellied Turtles Head Start Program
   Berkshire Museum and Aquarium
   Pittsfield, MA 01201
   413-443-7171
   www.berkshiremuseum.org

   Turtle Tots Program
   (Head start program for diamond-backed terrapins in Maryland)
   Maryland Department of Natural Resources
   410-260-8269
   www.dnr.state.md.us/fisheries/commercial/99ttots.htm

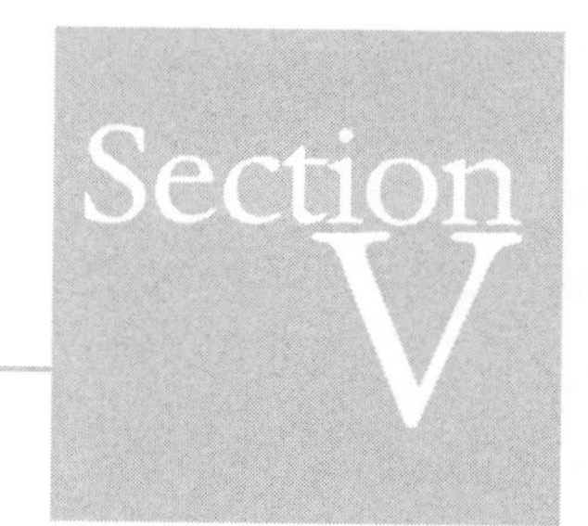

Herp Malformations and Declines: A Scientific Inquiry

Over the last three decades, scientists and conservationists became increasingly alarmed at the rate of declines and extinctions of amphibian species in the United States and other countries. Then, in the mid-1990s, a new concern grabbed the attention of the science and conservation communities—schoolchildren, naturalists, and scientists were finding an alarming number of malformed frogs. What could be causing these declines and malformations?

Through the activities in this section, students explore how scientists and volunteers are trying to solve these unhappy mysteries. Unlike the previous sections, where the emphasis is on understanding biological concepts and conservation as applied to herps, the activities in this section are designed primarily to enhance students' understanding of the scientific research process. Research on amphibian malformations and declines is used to illustrate the ways in which scientists and volunteers, including students, address environmental problems.

The activities in Chapters 17, 18, 20, and 21 are largely classroom discussions of issues related to conducting research on herp declines and malformations and other environmental problems. Activities in Chapter 19 involve students in conducting hands-on field research, including collaborating with other volunteer and professional scientists through the Internet. Although Chapters 17 and 18 provide interesting background information, students who want to do fieldwork can go straight to Chapter 19.

Chapter 17. Mystery of the Malformed Frogs

Chapter 18. Amphibian Population Declines

Chapter 19. Becoming Citizen Scientists: Surveying Amphibian Malformations and Declines

Chapter 20. Solving the Mystery of Amphibian Malformations

Chapter 21. Beyond Research: Scientific Disagreement, Ethics, and Policy

Chapter 17

Mystery of the Malformed Frogs

National Science Education Standards
Grades 5–8: Abilities necessary to do scientific inquiry (Science as Inquiry)
Understandings about scientific inquiry (Science as Inquiry)
Natural hazards (Science in Personal and Social Perspectives)
Science as human endeavor (History and Nature of Science)
Nature of science (History and Nature of Science)
Grades 9–12: Abilities necessary to do scientific inquiry (Science as Inquiry)
Understandings about scientific inquiry (Science as Inquiry)
Environmental quality (Science in Personal and Social Perspectives)
Natural and human-induced hazards (Science in Personal and Social Perspectives)
Science as human endeavor (History and Nature of Science)
Nature of scientific knowledge (History and Nature of Science)

The Discovery

On August 8, 1995, middle school teacher Cindy Reinitz took a group of students on a field trip to explore life in a local farm pond. What started out as a routine trip soon changed into an experience that would change the lives of the Minnesota New Country School students, and of scientists who study amphibians, forever.

A malformed frog.

The students noticed several frogs with missing legs. At first they thought maybe they had stepped on the frogs and injured them. But by the end of the day, 11 of the 22 frogs they had collected had extra or missing legs, or some other deformity. The students, and later scientists and the lay public, were faced with a dilemma. Is finding a population of frogs in which 50 percent of the animals are malformed a normal event, or does it point to a serious problem, perhaps with an environmental cause?

For the students, discovering the malformed frogs brought instant media attention. First, they were interviewed by a local newspaper. Then a TV station featured their discovery on the evening news. On August 25, 1995, the story hit the Associated Press and the school was inundated with phone calls from media across the country.

The discovery also brought several environmental and monetary awards to the school. Furthermore, the students testified in front of the Minnesota State Legislature in favor of a bill that would fund research to determine the causes of frog malformations.

Perhaps most interesting, the students became involved in a community of scientists, who did not have an answer to their questions: Is the incidence of malformed frogs increasing in North America? If so, what is causing the malformations?

Frog Malformations Research

Is the number of malformed frogs increasing? Scientists from the Minnesota Pollution Control Agency soon became involved in the study of malformed frogs. To determine whether the level of abnormalities observed at the Minnesota farm pond was a random event, or whether the incidence of malformed frogs was actually increasing, they needed frog surveys from more areas. By the end of 1996, the Minnesota Pollution Control Agency had received 175 reports of malformed frogs from two-thirds of Minnesota's counties. The agency also began to receive reports of malformations from other states.

However, some scientists countered the uproar about the recent reports with evidence that amphibian abnormalities have been observed since the 1700s. They believed that the current malformations observed in Minnesota and elsewhere were a normal or random occurrence. Despite these precautions, the majority of herpetologists now agree that the preponderance of recent reports is alarming and that we are probably seeing an increased incidence of frog malformations over what would normally be expected.

In response to the concern about amphibian malformations, the Biological Resources Division of the U.S. Geological Survey in the U.S. Department of Interior formed the North American Reporting Center for Amphibian Malformations (NARCAM). Scientists, as well as informed students and volunteers, can contribute to our knowledge about amphibian malformations by participating in this program.

What is the cause of amphibian malformations? The mystery of the malformed frogs is not yet solved, although many explanations have been proposed. To date, scientists have found evidence that may link malformations in tadpoles and frogs with a number of factors, including pesticides and other contaminants, ultraviolet radiation, and parasites. It is also possible that these factors interact to cause the malformations. For example, increased ultraviolet radiation caused by ozone depletion may transform nontoxic chemicals into toxins. Scientists at government agencies and universities continue to study the possible causes of malformations.

Sources

Minnesota Pollution Control Agency website *www.pca.state.mn.us/hot/frog-bg.html*

Minnesota New Country School Frog Project website *www.mncs.k12.mn.us/html/projects/frog/frog.html*

Souder, W. 2000. *A Plague of Frogs: The Horrifying True Story*. New York: Hyperion.

Activity 17.1. The Students' Discovery

Objective

Students will develop an understanding of how students and volunteers can make scientific observations that can aid conservation.

Materials

- computer with Internet access
- Minnesota New Country School Frog Project website *www.mncs.k12.mn.us/html/projects/frog/frog.html*
- copies of "Student Handout: Mystery of the Frogs" (provided)

Procedure

1. Have your students read the account of the Minnesota New Country School project by Ryan Fisher, "Mystery of the Frogs." Ask your students what they would have done if they had found that half of the frogs in a local pond were malformed. Would they have reported their results to anyone? If so, to whom? How might they know when their observations are important enough to report to a government agency or scientist?
2. Explain to the students that any population will have some malformed animals, and the number of malformed individuals will vary from one population to another. Ask your students how they might go about determining whether the malformations found at the Minnesota pond were a random event or an event that needed to be investigated.
3. Students may want to further explore the Minnesota New Country School Frog Project website, which includes an interactive page for monitoring herp malformations (see Materials, above).

Student Handout

Mystery of the Frogs

by Ryan Fisher, 10th Grade, Minnesota New Country School Frog Project

I've always been fascinated with science. So I jumped at the opportunity to get involved with the Minnesota New Country School Frog Project.

The frog project began on a summer nature trip led by Cindy Reinitz. Some students thought they had stepped on a few frogs. The frogs' legs appeared bent. I'm sure everyone must have checked the bottom of their shoes, as they tried to find the unknowing culprit. Soon the students realized that no one had stepped on the frogs. Puzzled by the mystery, everyone spread out, searching for more frogs. Over half of the frogs caught were malformed.

Cindy and her group returned to the pond several times in search of malformed frogs. Once the Minnesota Pollution Control Agency joined Cindy's group. They wanted malformed frogs to study. It was during those trips to the pond that Minnesota New Country School received the most publicity. A few TV stations broadcast live from the pond. Most of the publicity ended quickly. I guess the story of some students making a scientific discovery while sloshing through a pond lost its excitement after no answer was found immediately.

I became involved shortly after school began. Cindy Reinitz organized a meeting to plan the future of the frog project. It was at that meeting that we formed a battle plan of sorts. As we sat around Cindy in her classroom, we decided on how to attack this mystery that even the Minnesota Pollution Control Agency didn't have an answer for.

More trips to the pond were planned, water tests started, and students quickly volunteered to help with the different tasks on hand. From the beginning I worked on tracking down the leads we found. My problem was too much information. You have to be smart in deciding which lead could uncover that last elusive clue that would cause everything to make sense. That job required looking at science and the frogs in a different way. Instead of focusing on the obvious, I had to take the initiative to notice the details that might be overlooked. I had to find solutions on my own.

When others looked for someone to blame, I had to understand the importance of ethics and principles in finding the cure, not only the reason. Working on this project taught me how to do things I didn't think I could do. People probably will forget about the Minnesota New Country School Frog Project. But even as memories fade, one lesson won't be forgotten. A teacher and her students will have learned a lesson they never will forget: that if you try you will succeed, and you can make a difference in your town, your city, or even your country. Because you can't stop someone who thinks they can.

Source

Minnesota New Country School Frog Project
www.mncs.k12mn.us/html/projects/Frog/frog.html

Chapter 18

Amphibian Population Declines

National Science Education Standards
Grades 5–8: Abilities necessary to do scientific inquiry (Science as Inquiry)
Understandings about scientific inquiry (Science as Inquiry)
Natural hazards (Science in Personal and Social Perspectives)
Science as human endeavor (History and Nature of Science)
Nature of science (History and Nature of Science)
Grades 9–12: Abilities necessary to do scientific inquiry (Science as Inquiry)
Understandings about scientific inquiry (Science as Inquiry)
Environmental quality (Science in Personal and Social Perspectives)
Natural and human-induced hazards (Science in Personal and Social Perspectives)
Science as human endeavor (History and Nature of Science)
Nature of scientific knowledge (History and Nature of Science)

In the early 1970s, biologist Cynthia Carey was studying western toads in remote areas of Colorado. The first year she began her work, the toads were so numerous she had to be careful not to run over them with her car. Two years later, hardly any toads were left. Those still alive were barely able to walk and had puffy, red legs.

In the remote Sierra Nevada Mountains of Yosemite National Park, Gary Fellers admired frog choruses as a young boy and college student in the 1960s. By the 1990s, he found that many of the once resounding lakes were silent. Historical records dating from 1915 (such historical records are unusual for amphibians) showed that of seven species present shortly after the turn of the century, five had suffered serious declines by the 1990s. Only one species, the bullfrog, which is not native to the Sierras, was doing well.

In the early 1990s, similar reports of species declines and extinctions began pouring in from other corners of the globe. Costa Rican golden toads and harlequin frogs, plentiful in the early 1980s, disappeared suddenly. In Australia, the gastric brooding frog, the female of which swallowed its eggs and incubated them in her stomach, became extinct in 1980. These and other reports of extinctions prompted herpetologist Ronald Heyer to comment, "The frogs seem to be telling us something. We'd better find out what it is, and soon."

The story of frog declines has much in common with the story of amphibian malformations. Similar to the situation with malformations, controversy brewed as to whether the declines and extinctions were a new phenomenon or simply normal population fluctuations. A 1991 study of amphibians in South Carolina ponds showed that huge population fluctuations are indeed normal. The study also found that over many years, local extinctions (e.g., in a single pond) are also normal. However, many scientist believe that the number of recent extinctions and declines, and the fact that they are occurring on several continents, is alarming.

Concern over the recent population declines motivated scientists to look for a cause. Scientists have proposed a number of hypotheses to explain these declines, including the following:

Introduced species. To meet demands for recreational fishing, millions of fish have been stocked in lakes and rivers. Bullfrogs have also been introduced in many places in the western United States, where they are not native. These fish and bullfrogs may eat native amphibians or take over their habitat.

Human consumption. The Indian bullfrog may be experiencing declines due to the export of 3,000–4,000 metric tons of frog legs exported annually to France for human consumption.

Ultraviolet radiation. Field experiments in mountain lakes show that eggs of some frogs and toads are damaged by ultraviolet radiation. Scientists also found that a pathogenic fungus acts together with ultraviolet radiation to kill frog and toad eggs.

Acid rain and acid soil. Water and soil with pH below 4 can kill amphibian eggs and tadpoles. Field studies have shown that amphibians may disappear from regions that suffer from acid rain.

Pesticides. In Canada, scientists found only five species of frogs in lakes that had been sprayed with DDT and 12 species of frogs in lakes where no pesticides had been sprayed. DDT residues in frogs from sprayed lakes averaged 5,000–47,000 micrograms/kg, whereas frogs from unsprayed lakes had 6 micrograms/kg of DDT. In the Midwest, frog populations increased on a farm where pesticide use was stopped.

Disease. Scientists in Australia, Costa Rica, and the United States recently have discovered a new species of fungus in dead frogs. Evidence of trematode, bacterial, and viral infections also have been found.

Malformations. The observed high rates of malformations in some areas could play a role in population declines. Currently, none of the frog species with high rates of malformations is in danger of extinction, although malformations are suspected to be linked to amphibian declines.

Habitat destruction. Because amphibians spend time in both aquatic and terrestrial habitats, they suffer from loss of wetlands and forests. Some amphibians depend on temporary ponds that are present only in the spring (vernal ponds); these ponds are often drained for mosquito control, farming, or construction. Habitat loss probably is the single most important factor in the decline of amphibians and other animals worldwide.

Much of the concern over amphibian declines is based on the premise that amphibians may be especially sensitive to human disturbance and pollution in the following ways:

1. Amphibian eggs do not have protective shells and are extremely vulnerable to drying out.

2. Many amphibians lay their eggs in the water, where the eggs may be exposed to chemical pollution and ultraviolet radiation.
3. Introduced fish and other predators can easily prey on the aquatic eggs and young.
4. Many amphibians have highly permeable skin, which makes them susceptible to contaminants, both on land and in water.
5. Many amphibians depend on the preservation of both aquatic and terrestrial habitats to survive at different times in their lives. It is not enough to protect just one of these habitats.

People have a number of reasons for being concerned about malformations and population declines in amphibians. Some people feel that amphibians have value of their own, and that it is ethically wrong to harm them by contaminating the environment. Others may be more concerned with the possible meaning of amphibian malformations for humans. If amphibians are becoming unhealthy and dying, will humans soon follow? Are amphibians the "canary in the coal mine?"

Sources

Carey, C. 1993. Hypothesis Concerning the Causes of the Disappearance of Boreal Toads from the Mountains of Colorado. *Conservation Biology* 7 (2): 355–62.

Droege, S. An Outline of Issues Associated with Amphibian Declines: Brief Overview and History. North American Amphibian Monitoring Program (NAAMP). *www.mp1-pwrc.usgs.gov/amphib/frogsum.html*

Drost, C., and G. Fellers. 1996. Collapse of a Regional Frog Fauna in the Yosemite Area of the California Sierra Nevada. *Conservation Biology* 10 (2): 414–25.

Luoma, J. *Vanishing Frogs*. 1997. *Audubon* (May–June).

Activity 18.1. Why Care about Declines and Malformations?

Objective

Students will understand several arguments for why preservation of wildlife is important.

Procedure

1. Read your students the following quotes:

   "When extinctions occur among species whose roots on this planet surpass ours by millions of years, we should be listening to what they have to say." —scientist, North American Amphibian Monitoring Program

   "What's driving this whole issue is not deformed frogs. It's the potential for effects on human health." —toxicologist, Minnesota Department of Health

Do your students agree with the statements from the scientists? What do they think are important reasons for preserving frogs and other amphibians.?

2. Next, have your students brainstorm arguments for the preservation of endangered animals. Some reasons they may come up with are summarized below.

3. Have your students discuss or write an essay about the ethical issues related to amphibian malformations.

Reasons for Protecting Amphibians

Amphibians have a lot to teach us about science and biology. They are great examples of evolutionary success over a long period of time. They have thrived through 350 million years of dinosaurs, meteors, and humans. They have evolved fascinating adaptations to a wide range of environments, including oceans, deserts, ponds, and forests. They have efficient metabolisms. For example, about 95 percent of what a salamander eats becomes energy or gets stored as fat.

Amphibians play important roles in ecosystems. They are very abundant and often dominate ecosystems in terms of numbers and total weight, particularly in wetland and forest ecosystems. They are also integral parts of food webs. Amphibians are popular food items for many predators and important predators themselves.

Amphibians may be a measure of the health of the environment. It is possible that amphibian declines are a response to environmental pollution and degradation. Thus, amphibians may be showing us how our activities affect our shared biosphere.

Amphibians contain chemicals that may benefit humans. Amphibians have foul-tasting chemicals in their skin and glands that protect them from predators. Some of these chemicals can be used in medicine as, for example, heart stimulants, painkillers, and organ glues.

Amphibians can be used to control insect pests. In Australia, a tree frog is used to control insects.

Amphibians are fascinating and beautiful creatures. Many amphibians, such as tree and poison dart frogs, are beautiful beyond description. Many others are fascinating to children and adults. This fascination is illustrated by the role amphibians have played in literature, including children's stories, myths, the Bible, the Koran, and the works of Shakespeare.

Sources

Droege, S. *An Outline of Issues Associated with Amphibian Declines: Brief Overview and History.* North American Amphibian Monitoring Program (NAAMP). *www.mp1-pwrc.usgs.gov/amphib/fropgsum.html*

"Introduction to the Malformed Amphibian Issue." North American Reporting Center for Amphibian Malformations (NARCAM). *www.npwrc.usgs./gov/narcam*

Souder, W. 2000. *A Plague of Frogs: The Horrifying True Story.* New York: Hyperion.

"Why Do Amphibian Declines Matter?" Declining Amphibian Populations Task Force (DAPTF). *www.open.ac.uk/daptf/*

Chapter 19

Becoming Citizen Scientists: Surveying Amphibian Malformations and Declines

National Science Education Standards
Grades 5–8: Abilities necessary to do scientific inquiry (Science as Inquiry)
Understandings about scientific inquiry (Science as Inquiry)
Natural hazards (Science in Personal and Social Perspectives)
Science as human endeavor (History and Nature of Science)
Nature of science (History and Nature of Science)
Grades 9–12: Abilities necessary to do scientific inquiry (Science as Inquiry)
Understandings about scientific inquiry (Science as Inquiry)
Environmental quality (Science in Personal and Social Perspectives)
Natural and human-induced hazards (Science in Personal and Social Perspectives)
Science as human endeavor (History and Nature of Science)
Nature of scientific knowledge (History and Nature of Science)

Scientists studying the problem of amphibian malformations need more information on where malformations are occurring and how common they are. It is extremely difficult for scientists to conduct amphibian malformation surveys across large regions, and thus, many areas are not surveyed. By conducting careful observations in their own regions and then sharing these observations with scientists, knowledgeable students and volunteers can contribute greatly to solving the mystery of the malformed frogs.

The North American Reporting Center for Amphibian Malformations (NARCAM, part of the U.S. Geological Survey Biological Resources Division) maintains a database of sites where scientists and volunteers from all over the country have observed malformed amphibians. Students can contribute to this database by observing amphibians at their own field sites and recording their observations on forms found on the NARCAM website (*www.npwrc.usgs./gov/narcam*). To contribute to this effort, students will need to become familiar with different kinds of malformations, using the illustrations on the NARCAM website.

Through the NARCAM website, students also can access maps of their state that indicate where malformations have been observed, as well as sites where people have looked for malformations but have not found any. Students can report their own observations to NARCAM electronically over the Internet.

Similarly, through the North American Amphibian Monitoring Program (NAAMP, also part of the U.S. Geological Survey Biological Resources Division), scientists, students, and volunteers can contribute to our understanding about amphibian populations. Students and volunteers can use protocols published on the NAAMP website (*www.mp1-pwrc.usgs.gov/amphib/frogsum.html*) to survey populations of frogs, salamanders, and other amphibians. They then report their results to scientists through the website. Frogwatch USA is a sister program to NAAMP that is geared more toward youth. The NAAMP website has protocols for calling surveys and terrestrial salamander surveys, whereas the Frogwatch USA website (*www2.open.ac.uk/biology/froglog/*) has protocols for a simplified calling survey.

Calling Surveys

Many frog and toad species make unique calls during the breeding season that can be identified in the field. Calling surveys make use of these calls to track frog and toad populations (see Chapter 5). To participate in the Frogwatch USA or NAAMP Calling Surveys, your students will first need to learn how to identify these calls. A number of frog call tapes are available (see Frogwatch USA website [*www2.open.ac.uk/biology/froglog/*] and page 39 of this guide).

Once students have learned how to identify different frog species using their calls, students will need to become familiar with the calling survey sampling protocols that are available on the Frogwatch USA and NAAMP websites. The Frogwatch USA survey requires less time commitment than the NAAMP protocol; additionally, the NAAMP survey requires driving an assigned route in a car after dark, whereas students or volunteers in the Frogwatch USA survey choose their own nearby wetland.

Terrestrial Salamander Surveys

Because salamanders have not been well studied, we know little about their overall numbers and population trends. The Terrestrial Salamander Monitoring Program is designed to remedy this situation. This program seeks to answer the following questions: Are populations of terrestrial salamanders declining, similar to populations of frogs and toads? As eastern forests recover from widespread logging in the early 1900s, are populations of salamanders increasing?

Unlike many monitoring programs, the Terrestrial Salamander Monitoring Program provides opportunities for field experiments. In one experiment, participants determine the optimal time of year for sampling salamanders using boards as artificial cover objects under which the salamanders can be found. In another experiment, participants determine the preferences of salamander species for different configurations and types of cover objects. Most of the research on terrestrial salamanders is carried out in spring and fall, during the day, making it easy for students to participate. Protocols are available on the NAAMP website.

Other Surveys

Some states have undertaken comprehensive atlas projects to map the distribution of amphibians and reptiles across the entire state. Sometimes volunteers make observations that contribute to the atlas. Contact your state natural resources agency to determine the status of its herpetological atlas. Also, access state and federal herp websites to see what new surveys are being added and what surveys are being conducted in your area.

Activity 19.1. Monitoring Amphibian Malformations

Objective

Students will participate in an ongoing research effort to document the presence of malformed and normal amphibians.

Materials

- computer with Internet access
- NARCAM website (*www.npwrc.usgs.gov/narcam/index.htm*)

Procedure

1. Have your students access the NARCAM website. From the maps on the website, they will be able to see if volunteers or scientists are making observations of malformed herps in a nearby area. If just one person has access, he or she may want to print the maps of interest for the rest of the class.
2. Have students make their own local observations of amphibians using the guidelines for nonbiologists provided on the website. The guidelines will ask for the name of the species, whether or not malformations were observed, a description of the types of malformations observed, and information about habitat. An online species identification guide and description of the types of malformations that have been observed are included on the website.
3. It is essential that students and others participating in this program include "negative" observations (locations where they have found frogs without malformations) as well as "positive" observations (locations where they have found frogs with malformations). Can your students explain why it is important to report both negative and positive observations? (If participants in the survey only reported sites where they found malformed frogs, scientists would have no way of knowing how common malformed individuals are in relation to the total population of amphibians.)

Activity 19.2. Surveying Amphibian Populations

Objective

Students will participate in an ongoing research effort to document population trends of amphibians.

Materials

- computer with Internet access
- Frogwatch USA website (*www.mp2-pwrc.usgs.gov/frogwatch/index.htm*)
- NAAMP website (*www.mpl-pwrc.usgs.gov/amphib/frogsum.html*)

Procedure

1. Sampling amphibians presents some interesting challenges. Many salamanders, for example, spend most of their life under cover and are only seen during the short breeding season. Frogs may be high in trees, burrowed under rocks, or camouflaged by leaves right in front of your eyes. Ask your students how they would sample frog and salamander populations, based on what they have learned about the biology of these organisms through previous activities in this guide.
2. Have your students access the Frogwatch USA and NAAMP websites (see Materials, above) and learn about the different amphibian monitoring programs. The Frogwatch USA monitoring program is less rigorous, and thus may be more appropriate for students.
3. Decide as a group or as individuals which monitoring programs you would like to participate in. Follow the procedures on the websites to carry out the protocols related to the studies you have chosen.

Source

Mitchell, J. C. 2000. *Amphibian Monitoring Methods and Field Guide*. Front Royal, VA: Conservation Research Center, Smithsonian National Zoological Park.

Chapter 20

Solving the Mystery of Amphibian Malformations

National Science Education Standards
Grades 5–8: Abilities necessary to do scientific inquiry (Science as Inquiry)
Understandings about scientific inquiry (Science as Inquiry)
Natural hazards (Science in Personal and Social Perspectives)
Science as human endeavor (History and Nature of Science)
Nature of science (History and Nature of Science)
Grades 9–12: Abilities necessary to do scientific inquiry (Science as Inquiry)
Understandings about scientific inquiry (Science as Inquiry)
Environmental quality (Science in Personal and Social Perspectives)
Natural and human-induced hazards (Science in Personal and Social Perspectives)
Science as human endeavor (History and Nature of Science)
Nature of scientific knowledge (History and Nature of Science)

What are the causes of amphibian malformations? The proposed answers to this important question are fraught with controversy. In fact, the answers to many environmental questions, such as "Does acid rain damage forests?" or "What causes global warming?" are controversial. Why are questions having to do with environmental phenomena so difficult to answer?

The public often views science as a series of prescribed steps that result in definitive answers to proposed questions. In reality, however, it is not possible to directly answer some scientific questions. For example, we can test the effect of a potential contaminant on a mouse, but it is difficult to translate the results of these controlled laboratory studies directly into knowledge about how the particular chemical may affect humans. Because it is not always possible to test humans, issues related to human health often are complex and answers are rarely straightforward. Similarly, it often is impossible to come up with a single test to answer environmental questions. Therefore, environmental scientists conduct a number of different types of studies to answer a single question. Although no one study may be conclusive by itself, each one offers some information toward solving the problem. Scientists piece together evidence from these multiple studies and then suggest an answer to a question, such as, "What is causing amphibian malformations?"

Activity 20.1. Environmental Research

Objectives

Students will develop an understanding of (a) potential causes of amphibian malformations, (b) difficulties involved in answering environmental questions and why scientific controversy exists, and (c) different types of environmental research.

Materials

- copies of "Student Handout: Piecing Together the Puzzle: Types of Environmental Studies" (provided)

Procedure

1. Ask your students to brainstorm about possible environmental causes for amphibian malformations. They may come up with reasons similar to those scientists have proposed, including the following: pesticides, ultraviolet radiation, disease, acid rain, heavy metals, and predation.
2. Next ask them to propose a study to test for these possibilities. They may suggest one or more of the types of studies outlined in the student handout "Piecing Together the Puzzle: Types of Environmental Studies."
3. Once your students have proposed several different types of studies for testing their hypotheses about the causes of malformations, ask them to consider some of the problems and advantages of the various studies. For each type of study, ask students to suggest potential logistical problems with conducting the study, problems with interpreting data, and the value of each type of study. You can guide the students' discussions using the student handout.
4. As an extension of this activity, assign small groups of students to each type of study on the handout. Ask students to find other examples of their particular type of study on the Internet or in other readings. Then hold a class debate, in which groups of students present arguments for the importance of their particular study.
5. As a final step, engage your students in a discussion of why it is important to have several different kinds of evidence when answering environmental questions. Ask your students to draw a diagram of different types of studies they might undertake to answer the mystery of the malformed frogs. Students may want to conduct further research on the Web to find out what steps researchers have taken to solve this mystery.

Student Handout

Piecing Together the Puzzle: Types of Environmental Studies

Many scientific studies start with a simple observation, like the observation of malformed frogs made by the Minnesota New Country School students on a pond field trip. The students' observations raised many questions: "Is this a chance occurrence or an indication of a serious environmental problem?" "Are there high percentages of malformed frogs at other sites?" "What is the cause of the malformations?"

It is not possible to design one study to answer these questions. Rather scientists piece together an answer based on the results of many different types of studies such as the ones described below.

MONITORING STUDIES

What Is Monitoring?

In monitoring studies, professional and amateur scientists make observations at a wide variety of sites to see if there are any trends over time or patterns in space.

Logistics

Monitoring studies require observations from many sites over a large geographic area, and over a long period of time. For this reason, they are difficult, time intensive, and expensive for scientists to carry out. One solution to this problem is involving students and volunteers in collecting the data. This is the idea behind the North American Reporting Center for Amphibian Malformations (NARCAM) monitoring project, as well as a number of other monitoring programs for fish, amphibian, and bird populations.

Interpreting Results

Monitoring studies can indicate trends, and if historical data are available, they can also indicate where there might be potential problems. Unfortunately, for most herps, we do not have good historical data. Therefore, we do not know normal rates of malformations or even normal population levels. If 10 percent of the frogs in a population are malformed, how does this compare with normal expectations? In the absence of historical data, we can only guess. For this reason, disagreements about interpretation of a monitoring study's results are common. In addition, although monitoring studies might suggest a potential problem, they are not well suited for determining the cause of the problem. Even if monitoring studies show higher levels of malformations near sites of known contamination, we have no proof that the contamination actually caused the malformations. The malformations may have been caused by some other factor, such as disease.

Value

Despite the difficulties in interpreting monitoring studies, they are useful as a warning of impending problems. Monitoring studies also can be used to direct conservation management activities and funding, and to generate hypotheses for further research. Without monitoring programs, animal populations may dwindle without even a notice. Or before we become aware, populations can collapse to the point that conservation and management options become limited, expensive, or even impossible. This is what has happened with a number of species of amphibians worldwide.

Example

Sites throughout North America are being monitored to determine whether the incidence of malformations is increasing. This type of monitoring study was the first step scientists took after receiving information from the Minnesota New Country School students. Annual calling frog surveys are another example of monitoring.

Source: NARCAM (*www.npwrc.usgs.gov/narcam/index.htm*)

FIELD STUDIES

What Are Field Studies?

In field studies, scientists find sites that differ in some known way and make observations and take measurements at the different sites.

Logistics

Scientists conducting field studies need to find sites that are similar enough to allow them to make conclusions. For example, if a scientist were looking at amphibian populations in lakes with and without contaminants, she or he would want to be sure that the lakes were similar in size, temperature, predator populations, and other factors that might affect amphibians. Ideally, the only difference between the study sites would be the presence of contaminants. In reality, most often it is difficult to find areas that are similar in every way, except for the one factor scientists are interested in. To overcome little differences, field studies should be conducted in different seasons and over several years, to account for variability in uncontrolled factors, such as weather and food availability, that could affect results.

Interpreting Results

Scientists might determine that ponds near areas where pesticides have been applied have higher rates of amphibian malformations. However, this association does not prove that the pesticides caused the malformations. For example, it could be that these ponds also have higher nitrate levels because of run-off from manure. Or, frogs could be suffering from diseases that only occur around farms. In other words, just because an association exists between a contaminant and malformations, it does not mean the contaminant caused the malformations. Another word for association is correlation. Scientists often say, "Correlation does not equal causation."

Value

Field studies that uncover associations do point to possibilities that could be tested in a more controlled setting. However, scientists may not have the resources to conduct more controlled studies, and policymakers may not have the time to wait for more definitive results. In the absence of more concrete proof, careful field studies may provide reasonable evidence to help improve management plans and conservation actions.

Example

In a field study in Quebec, Dr. Martin Ouellet and colleagues examined over 1,000 frogs and toads from ponds and ditches subject to pesticide applications and from similar habitats free of agricultural contaminants. In sites with no pesticide use, 0.7 percent of the frogs had malformations. But in sites where pesticides had been sprayed, 12 percent of the frogs had malformations. However, there was a great deal of variability within the two types of sites. Some agricultural sites had 0 percent malformed amphibians, and at the most extreme site, 69 percent of the frogs and toads had malformations; the range in nonagricultural sites was 0–7.7 percent malformed amphibians. Dr. Ouellet and colleagues concluded that agricultural run-off was a likely cause of amphibian malformations but that a larger number of sites should be sampled.

Source: Ouellet, M., J. Bonin, J. Rodrigue, J.-L. DesGranges, and S. Lair. 1997. Hindlimb Deformities (Ectromelia, Ectrodactyly) in Free-Living Anurans from Agricultural Habitats. *Journal of Wildlife Diseases* 33: 95–104.

LABORATORY STUDIES

What Are Laboratory Studies?

Laboratory studies involve artificially imposing different levels of a treatment on a test subject (e.g., low and high levels of pesticides) and measuring a response (e.g., reproductive deficiency or number of malformations). These studies are conducted in a laboratory where the researcher can control variables, such as light and temperature, that might affect the results.

Logistics

Laboratory studies eliminate problems with travel, vagaries of weather, and uncontrolled disturbances that occur with field studies. Therefore, it is common to bring study animals into the controlled environment of the lab to sort out intricate problems. However, lab studies with animals face the difficulty of keeping animals healthy and alive in the laboratory. Additionally, there may be predicaments about choosing appropriate test organisms. Decisions must be made about which species to examine, and at what stages. Adults already have a history of behavior and exposure to the environment at their site, often making them undesirable for lab observations. If eggs were collected from the field, how would the scientists know whether or not they already are contaminated? Finally, there often are some ethical questions associated with removing animals from the wild, or subjecting animals to laboratory tests.

Interpreting Results

Conditions in the laboratory can never exactly duplicate the complex conditions in the field. Even if we can show that a pesticide causes defects in the laboratory, we can't be sure it causes defects under normal environmental conditions. On the other hand, if a lab study turns out negative results, can we conclude that there is no effect of the pesticide? It might be possible that the pesticide breaks down into a toxic by-product in the presence of ultraviolet light in a pond, but not under laboratory lighting. The most effective laboratory studies generally deal with only one factor, or a limited number of factors, to simplify the interpretation of results. In addition, conclusions always take into account that under real-life environmental conditions, many more factors can interact to complicate things, including sunlight, temperature, moisture, pollutants, and even other organisms.

Value

Positive results from a carefully conducted laboratory study could add strong evidence, which along with evidence from a field study, may point to a possible cause for malformations. A good lab study will help provide direction for future field studies.

Example

Dr. James La Clair and other scientists tested the effects of pesticides on early amphibian development. They found that S-methoprene, an insecticide used in the late 1970s to control fleas and mosquitoes, did not affect the development of frog embryos. However, when exposed to sunlight, water, and microorganisms, such as one might find in the environment, S-methoprene broke down into products that dramatically altered the development of the embryos. The scientists found that the embryos developed into juveniles with malformations similar to those found in nature. La Clair concluded that the current procedure of assessing the risk posed by pesticides by examining only the pesticide is not sufficient. Rather, it is crucial to include the relationship between amphibian development and the degradation products of pesticides under natural conditions.

Source: La Clair, J., J. Bantle, and J. Dumont. 1998. Photoproducts and Metabolites of a Common Insect Growth Regulator Produce Developmental Deformities in *Xenopus*. *Environmental Science & Technology* 32 (10): 1453-61.

BIOASSAYS

What Are Bioassays?

Bioassays are a type of laboratory study in which a standard test subject, such as lettuce seeds, daphnia, or frogs, is subject to a potential contaminant and monitored closely. Conditions that might affect the results, such as length of time the organism is exposed to the contaminant, temperature, and light, are tightly controlled throughout the study. Scientists then observe health, growth, abnormalities, and mortality of the organism.

Logistics

Bioassays generally are simple, quick, and inexpensive to conduct. However, scientists must pay strict attention to controlling environmental conditions, usually following a standardized method that is well established and accepted by others using the assay.

Interpreting Results

Although results from bioassays may point to a problem with water quality, they do not necessarily identify the chemical(s) that are causing the problem. Alternatively, no measured affect in a bioassay experiment does not mean that there are no toxins present. Additionally, bioassays only measure the test organism, which may not behave the same as the organisms that are of concern in the field. A common test subject for amphibian toxicity studies is the African clawed frog, which is not native to the United States.

Value

Bioassays often are used to identify hot spots of pollution, which scientists then may want to investigate more thoroughly using chemical tests.

Example

Scientists from the National Institute of Environmental Health Sciences tested water samples from sites at which frog malformations were found and from sites where no malformed frogs were found. They used a test called a FETAX assay, which is a standard bioassay used to test for the presence of toxins in water. In this test, eggs of African clawed frogs are grown in water, and abnormalities and mortality are recorded. The tests were run multiple times, using dilutions of the water from the study sites ranging from 0 to 100 percent. At concentrations above 50 percent of the water from the sites where there were frog malformations, a high percentage of the frog embryos showed abnormalities similar to what has been observed in tadpoles in the field. In contrast, water from sites where no malformed frogs were found did not produce harmful effects in the tadpoles in the bioassay. Alarmingly, the FETAX test gave positive results using both surface water and groundwater from the affected sites, including tap water from private wells used by the closest house to each site.

Minnesota Pollution Control Agency Commissioner Peder Larson concluded, "These findings give us a big piece of the puzzle we've been looking for in regard to the problems with the frogs. It does not provide evidence of a human health link, but it does underline the need to look more closely at what all this may mean for the environment. If the frog investigation was a priority for us before, it's even more so now."

Source: Minnesota Pollution Control Agency (*www.pca.state.mn.us/hot/frog-bg.html*)

CHEMICAL ANALYSES

What Are Chemical Analyses?

Chemical analyses are used to identify specific contaminants in water, soil, or plant and animal tissues. These tests often are used to determine what chemicals are present in a sample of water or a soil sample that has been shown to adversely affect organisms in a bioassay.

Logistics

Because there are so many potential contaminants, these tests can be time consuming. In addition, they often require expensive supplies and equipment. Some analyses can only be done in qualified laboratories.

Interpreting Results

Tests can show a particular chemical was in the water but cannot determine what the effect of that chemical was under the conditions where the frogs were found.

Value

In cases where the effects of a chemical on organisms are known, information about the presence of that chemical is essential in solving environmental problems. Chemical tests also can provide important clues for further studies.

Example

After discovering that well water caused frog malformations, the Minnesota Pollution Control Agency tested the water for chemicals. They found several pesticides and fungicides in the water. (Subsequent laboratory studies showed that the pesticides and fungicides cause limb malformations in frogs.)

Source: Minnesota Pollution Control Agency (*www.pca.state.mn.us/hot/frogs.html*)

CONTROLLED FIELD EXPERIMENTS

What Are Controlled Field Experiments?

Controlled field experiments involve artificially imposing different levels of a treatment under carefully planned field conditions.

Logistics

Controlled field experiments are perhaps the most difficult logistically. Can you find two sites that are exactly the same? Can you get permission to apply pesticides or disease agents? What are the ethics of applying chemicals that you know have negative effects on organisms and their environment? What are the ethics of intentionally introducing a suspected pest or disease into the wild? How many replicate sites do you need? What if rainfall or temperature is above normal the year the experiment is conducted?

Interpreting Results

Although controlled field experiments simulate natural processes in the environment, it may be impossible to exactly duplicate natural conditions that occur over a long time period. For example, pesticide residues may build up over a number of years, and so the control sites may not be completely free of pesticides. It is always difficult to interpret the results of a study site without taking into account the often unique past history of the site.

Value

Controlled field experiments more accurately mimic natural conditions than laboratory research, yet have more stringent controls than field studies. If conducted properly, they therefore provide the strongest evidence of a link between a cause, such as a chemical, and an environmental problem, such as disease, in the field.

Example

Scientists could apply pesticides to several ponds in the field while leaving others unaltered (as a control) and then measure malformations in both types of ponds.

Chapter 21

Beyond Research: Scientific Disagreement, Ethics, and Policy

National Science Education Standards
Grades 5–8: Abilities necessary to do scientific inquiry (Science as Inquiry)
Understandings about scientific inquiry (Science as Inquiry)
Natural hazards (Science in Personal and Social Perspectives)
Science as human endeavor (History and Nature of Science)
Nature of science (History and Nature of Science)
Grades 9–12: Abilities necessary to do scientific inquiry (Science as Inquiry)
Understandings about scientific inquiry (Science as Inquiry)
Environmental quality (Science in Personal and Social Perspectives)
Natural and human-induced hazards (Science in Personal and Social Perspectives)
Science as human endeavor (History and Nature of Science)
Nature of scientific knowledge (History and Nature of Science)

In Activity 20.1, we saw that no one study or experiment is likely to provide a definitive answer to complex environmental questions such as causes of amphibian malformations. Rather, scientists must piece together evidence from a variety of studies, including field observations and lab and field experiments, to reach a reasonable conclusion. Because each of the studies that are pieced together will have some uncertainty, there almost always will be some scientists who disagree with the conclusion of the majority of scientists.

This disagreement among scientists has both negative and positive consequences. On the negative side, strong disagreements may hinder productive debate, and thus prevent scientists from working toward a common goal of understanding. Also, politicians and the media often pick up on scientific disagreement, and may use the results of the minority group of scientists to back a particular point of view. An often-used argument is that "even scientists don't agree." This statement can be used to exaggerate the degree of disagreement among scientists and to justify not taking action on a large problem. In many cases, scientists may disagree on small points but are in general agreement on larger issues. For example, the vast majority of scientists now agree that global warming is occurring, yet they may disagree on the degree of global warming and its effects on ecosystems. Nevertheless, many politicians contend that even scientists disagree about whether global warming is occurring.

Disagreement also can be a positive force in science. It often sparks productive debate among scientists, leading to creative new ideas and solutions. It also encourages others to join in the search for answers to scientific questions. Healthy disagreement keeps us all from being lazy and settling on a single (and possibly wrong) answer.

The difficulty of finding scientific answers to environmental questions not only causes disagreements among scientists, it also makes it difficult to translate research results to environmental policy. Therefore, policymakers nearly always make decisions without complete scientific information.

Along with the inherent uncertainty, scientists studying environmental problems also often face ethical questions. When researchers at the Minnesota Pollution Control Agency found that water from wells caused malformations in African clawed frogs, what was their responsibility to the residents who used these wells? Is it the right thing to do to warn people that their water may be contaminated, even if you are not sure?

Activity 21.1. Scientific Disagreement

Objective

Students will gain an awareness of disagreement among scientists.

Materials

- copies of "Student Handout: Scientific Disagreement" (provided)

Procedure

1. Have your students read the handout "Scientific Disagreement." Ask them to discuss why this type of disagreement among scientists is common.
2. Can your students think of any reasons why this kind of disagreement is good or bad? Can they think of any current or historical examples of disagreements among scientists?

Student Handout

Scientific Disagreement

"Scientists' Only Consensus Is Deformed Frogs Remain a Mystery"
by Tom Meersman, *Minneapolis Star Tribune* staff writer

November 21, 1997

SAN FRANCISCO—Scientists at a national conference Thursday came up with plenty of ideas about what might be causing frog deformities in Minnesota and other states, ranging from ultraviolet radiation and pesticides to leaking landfills and even tadpole cannibalism.

About the only thing researchers could agree upon during a four-hour session was that eye, limb and internal deformities documented since 1995 are still a mystery and probably will prove to have several causes.

More than one compound can cause some malformations, said Jack Bantle, a zoology professor at the University of Oklahoma. "You may have different answers regionally, and you shouldn't just look for one answer."

He is one of several scientists who has been examining methoprene, a relatively common pesticide used to control mosquitoes and a few other insects. Some researchers, including those at the Environmental Protection Agency lab in Duluth, have exposed frog eggs to methoprene and found no abnormalities.

But Bantle said he has found that ordinary sunlight breaks methoprene down easily and quickly into chemical byproducts—some of which last much longer than methoprene does and appear to be more harmful to frogs in their early stages.

Bantle presented evidence that some methoprene byproducts, when added to fertilized frog eggs in a lab, produced a host of deformities in tadpoles, including head and eye malformations, facial defects and heart problems. "You have to take this data with a grain of salt," he said, because no one knows whether what has been tested in the lab also can be found in frog breeding ponds in Minnesota or elsewhere.

Water a Factor in Debate

David Hoppe, biology chairman at the University of Minnesota-Morris, said that he continues to find high percentages of deformities in such species as the green, mink and northern leopard frogs that spend much of their lives in water. "I've seen at least 80 mink frogs with grotesque deformities this year."

The more terrestrial species in Minnesota—such as the gray tree frog, American toad, wood frog and spring peeper—also show deformities, he said, but they appear to be less vulnerable.

Minnesota Pollution Control Agency (MPCA) officials released new figures at the conference for the number of frog deformities reported during 1997. Judy Helgen, MPCA wetlands biologist, said more than 200 citizen reports were filed with the agency (about the same as in 1996), and scientists have confirmed abnormal frogs at 62 sites in 29 counties, not including some of the areas Hoppe has been using.

Last month the MPCA reported that water from four wells near ponds with deformed frogs had induced deformities in fertilized frog eggs in lab tests. Helgen said Thursday that 34 more wells have been sampled near other sites, and that lab-test results with that water should be available within a couple of weeks.

Environmental Protection Agency officials have criticized the methodology of that study, which was conducted through the North Carolina-based National Institute of Environmental Health Sciences, the MPCA's federal partner in the project.

Jim Burkhart, the institute's lead scientist in the research, said extensive data back the preliminary results that suggest something in the water may cause some of the deformities.

He said the frog egg test has been blown out of proportion and is only a tiny part of the larger project that is analyzing hundreds of Minnesota water samples for dozens of chemicals, metals and other compounds. He said it will take several months to complete the analysis.

Stan Sessions, a biology professor at Hartwick College in New York, disagreed with many of the other researchers and said that natural phenomena can explain the abnormalities. Parasites that burrow into tadpoles and disrupt their normal limb development are responsible for multi-legged abnormalities, he said, and the missing limbs, eyes and other oddities probably were caused by tadpoles "chewing on each other," which he said he has observed in his lab.

Research will continue in several directions. The institute is moving to involve scientists from other disciplines, such as human health, in the discussion. A workshop next month will focus on potential human health concerns.

Activity 21.2. Science and Policy

Objective

Students will understand that policymakers often must make decisions with incomplete scientific information.

Procedure

1. Read your students the following:

> From his childhood days in the 1960s until the 1990s, ecologist Gary Fellers observed severe declines or extinctions in seven species of frogs and toads in lakes in Yosemite National Park. In a 1991 article in *Science*, researchers in South Carolina showed that extreme population fluctuations are normal in salamanders and frogs. Fellers offered the following comment on individuals who question frog declines, whom he refers to as "naysayers."
>
> "The naysayers can always say you won't know if you're seeing a trend until you survey these animals for a hundred years. But if you go to Yosemite and visit every pond, lake, and stream that is suitable for frogs and find major declines—or absolutely nothing—where there once was abundance, that's well beyond the scope of a natural fluctuation." *Audubon* (May-June 1997).

2. Ask your students if they agree that individuals who don't think we should act now are naysayers.
3. Ask your students to discuss the following question: If you were working for the EPA and charged with setting policies on wetlands and contaminants, what kind of evidence would convince you that frog populations were declining?
4. As part of the discussion, point out that sometimes political pressure or concern about the well-being of our planet forces policymakers to act without all the scientific facts.

Activity 21.3. Ethics and Policy

Objective

Students will become aware of an example ethical issue that policymakers must address.

Materials

- copies of "Student Handout: Piecing Together the Puzzle: Types of Environmental Studies"(provided)

Procedure

1. Have your students read the description of the FETAX bioassay on page 119 of the student handout.

2. Discuss with your students the following questions: Should the scientists report these results to the residents? Does the government have a responsibility to provide bottled water to the residents? What should the government scientists tell the residents about the risk to human health?

Activity 21.4. Herp Websites

Objective

Students will learn to use the Web to research amphibian natural history, identification, and conservation.

Materials

- computer with Internet access
- copies of "Student Handout: Herp Websites" (provided)

Procedure

1. Have your students access the different amphibian websites. They will likely find additional websites, which they can share with the group.
2. Have the students compile reports about a herpetological question of interest to them, and share information they learned from the websites.
3. Ask your students whether there is a way to evaluate the quality of the information they read on the Internet. How might they check up on information they find on the Internet to see how accurate it is? (For example, they might communicate with a herpetologist at a state agency or university or read an original article written by a scientist.)
4. Students may want to compile a new list of websites and share it with other classes or out-of-school groups interested in conservation of herps.

Student Handout

Herp Websites

Much of the information for Section V of *Hands-On Herpetology* was gleaned from the numerous amphibian websites. Students who want to learn more about amphibians can explore the websites listed below. Included are sites specifically focused on malformations and declines, as well as on education, conservation, and research. Many sites have illustrations of amphibians.

Declining Amphibian Populations Task Force (DAPTF)

www.open.ac.uk/daptf/index.htm

The mission of the DAPTF is to determine the nature, extent, and causes of declines of amphibians throughout the world, and to promote means by which declines can be halted or reversed. The group is based at the Open University Ecology and Conservation Research Group in Great Britain, and operates under the umbrella of the International Union for the Conservation of Nature Species Survival Commission. Established in 1991, the DAPTF consists of a network of over 3,000 scientists and conservationists from 90 countries. The site includes scientific and conservation information, links to other amphibian sites, and *FROGLOG* (*www2.open.ac.uk/biology/froglog/*), which has articles on many aspects of amphibian declines.

Frogwatch USA

www.mp2-pwrc.usgs.gov/frogwatch

Frogwatch USA relies on volunteers to collect information regarding frog and toad populations in neighborhoods across the nation. The website includes user-friendly background information and protocols for using breeding calls to monitor frogs and toads in wetlands.

Minnesota New Country School Frog Project

www.mncs.k12.mn.us/html/projects/frog/frog.html

Students at the Minnesota New Country School made the discovery of malformed frogs that brought national attention to the issue of frog malformations. Students at the school designed this website, which details their activities.

Minnesota Pollution Control Agency

www.pca.state.mn.us/hot/frogs.html

This site includes background information about malformed frogs, fact sheets on potential causes for malformations, and live photos of malformed frogs being cared for by the Minnesota Pollution Control Agency.

North American Amphibian Monitoring Program (NAAMP)

www.mp1-pwrc.usgs.gov/amphibs.html

NAAMP is part of the U.S. Geological Survey Biological Resources Program and is the North American component of the international Declining Amphibian Populations Task Force (see above). Its mission is to research causes of declines and to develop means to conserve amphibian species in Canada, the United States, and Mexico. Background information and instructions for monitoring amphibians are included on the site, including calling surveys, terrestrial salamander monitoring, aquatic surveys, atlas projects, and western surveys. Volunteers can contribute data to the calling surveys and terrestrial salamanders monitoring program. The data become part of the global pool of information being used to understand why amphibians are disappearing and how we can save them. The NAAMP site also includes the *Teachers' Toolbox* (*www.mp1-pwrc.usgs.gov/amphib/tools/teachers.html*), which provides user-friendly amphibian-monitoring and -surveying activities.

North American Reporting Center for Amphibian Malformations (NARCAM)

www.npwrc.usgs.gov/narcam/

NARCAM is part of the U.S. Geological Survey Biological Resources Program. Its mission is to facilitate the transfer of information on malformed amphibians. By compiling information from both the public and the scientific community, NARCAM hopes to convey accurate information on amphibian malformations and to encourage collaboration among scientists working to understand the cause(s) of amphibian malformations in the wild. The NARCAM site includes background information, illustrations, and monitoring procedures for malformed frogs.

Partners in Amphibian and Reptile Conservation (PARC)

www.parcplace.org

The mission of PARC is to conserve amphibians, reptiles, and their habitats as integral parts of our ecosystem and culture through proactive and coordinated public/private partnerships. PARC hopes to change public and political attitudes through educational efforts to raise public awareness and by promoting sound conservation and management strategies for our native reptiles and amphibians. It strives to include all individuals, organizations, and agencies with an interest in amphibian and reptile conservation, including representatives from federal and state agencies, conservation organizations, museums, nature centers, universities, research laboratories, the forest products industry, the pet trade industry, and environmental consultants and contractors. PARC's website includes educational activities for grades K–4, species accounts, and other educational resources.

A Thousand Friends of Frogs

www.cgee.hamline.edu/frogs/index.html

A Thousand Friends of Frogs connects students, educators, families, and scientists in order to study and celebrate frogs and their habitats. The site includes science, art, and language arts activities and resources.

Conclusion

Finding the causes for amphibian declines and malformations is an ongoing scientific effort. As the body of research grows, scientists and the general public will eventually come to a general consensus on causes of amphibian declines and malformations, although we can still expect some disagreement among researchers. Students and volunteers can contribute to these important research efforts by participating in the monitoring programs mentioned in this section. By doing so, they will be contributing to our understanding about the health of these unique organisms, and the health of our planet.

Appendix

Herp Species Accounts

By Martin A. Schlaepfer and Stephen J. Morreale

Reptiles and amphibians—or "herps"—have a universal appeal. Young students, especially, find it exciting to encounter them in the wild and, in some cases, to hold a herp in their hands. Although most encounters are short in duration, a deep satisfaction comes when a student is able to properly identify the animal and associate it with interesting facts about where it lives, what it eats, and how it functions. This type of information, which is considered natural history, reflects the uniqueness of each species.

The natural history information presented in the following four species accounts represents a summary of years of study by numerous biologists. These four accounts are a subset of a larger package being developed and written by Martin A. Schlaepfer and Stephen J. Morreale. The animal illustrations are by Abigail Rorer and are reprinted with permission from *Amphibians and Reptiles of New England* by Richard DeGraaf and Deborah D. Rudis (Amherst: University of Massachusetts Press, 1983) and *New England Wildlife: Habitat, Natural History, and Distribution*, by Richard M. DeGraaf and Mariko Yamasaki (Hanover, NH: University Press of New England, 2001).

Species Name: Spotted salamander *Ambystoma maculatum*

Description: The spotted salamander is large and stout, with a broad, blunt head. It is recognized easily by the round yellow or orange spots on its back, arranged in two irregular rows running down the length of its black or dark gray body. There can be as many as 50 spots, and these usually extend from the head to the tip of the tail. The belly tends to be a slate-gray color with gray flecks along the sides. Adults generally measure 11–19 cm (4–7 in), and can be as large as 25 cm (10 in). The aquatic larvae of spotted salamanders are dull green with white or light bellies, and generally lack any particular markings.

Similar species: The eastern tiger salamander sometimes has yellow dots, but these usually are irregular in appearance and are not arranged in two rows. The tiger salamander also has a yellowish belly. In the northeastern region, the tiger salamander is found only along the coast, and only as far north as Long Island.

Habitat: The preferred habitat of spotted salamanders is a mature deciduous forest with soft, moist soils and a nearby temporary or semi-permanent pond. They also occur in bottomlands and flood plains and can be encountered in mixed or coniferous forests. These salamanders and their relatives are burrowing animals that spend much of their time underground, or buried deep inside logs, except during periods of breeding activity.

Where and when: The spotted salamander is relatively common and widespread in eastern North America, extending from Nova Scotia to southeastern Ontario, southward to Georgia, and as far west as eastern Texas. Spotted salamanders are most noticeable in the early spring when they congregate in large numbers to breed over a short period of time. During this period of explosive breeding, which usually occurs in March or early April in the North, spotted salamanders can be seen making mass migrations toward nearby pools and ponds. The breeding migration generally is triggered by the first warm, steady spring rains, even if there is snow remaining on the ground. The males, who often arrive first, begin swimming about in a highly active state that becomes nearly a frenzy when females arrive in the pond to mate. This activity can be seen readily, even by a casual observer. During the rest of the year, the spotted salamander is largely secretive, retreating to underground burrows. In moist environments or damp weather, individuals can be encountered under logs, stones, or boards during the day, or out foraging at night. In winter they hibernate underground in burrows sometimes more than 1 m deep.

Natural history: During courtship and mating, adult male spotted salamanders deposit gelatinous sperm packets on sticks or on the bottom of the pond. The female then swims over the packet and takes up the sperm into her cloaca. Within one to a few days, the female lays eggs in gelatinous masses of usually 100 to 200 eggs. The egg clusters are attached to aquatic vegetation or sticks, surrounding the attachment site. Eggs usually take from 30 to 50 days to hatch, depending on the temperature of the water. The new hatchling starts out as an elongate larva, with gills near its neck region, and short buds in place of front limbs. As the larva develops, toes form on the front feet, rear legs sprout near the base of the tail, and it ultimately loses its gills and tail fin, all in preparation for life on land. Temperature, water level, and food availability combine to influence the length of the tadpole stage. The minimum time it takes for a spotted salamander to metamorphose into its terrestrial form is two months; usually newly transformed animals begin leaving the water in late summer and early fall. In the water, the larvae eat small crustaceans, mollusks, and insect larvae. On land, spotted salamanders eat beetles, earthworms, snails, slugs, insects, and spiders. Once transformed, they will remain on land for the rest of their lives, except briefly during breeding periods. Males reach maturity usually when they are two to three years old, whereas females take usually one to two years longer. A spotted salamander can live for more than 20 years.

Conservation status: The spotted salamander is relatively abundant in many habitats where it occurs throughout eastern North America. It is an important component in both aquatic and terrestrial communities. Eggs and larvae provide food for a wide variety of aquatic animals, and adults are eaten by predatory fish, birds, snakes, and turtles. There is some concern about potential decline of this species, due mainly to human pressures. In New York, it was shown that eggs can experience high mortality from lower pH levels due to acid precipitation. In addition, because of their complex habitat requirements, spotted salamanders are sensitive to the loss of both wooded and aquatic habitats. Furthermore, their tendency to migrate between these habitats during the breeding season makes them highly vulnerable to mass mortality. Substantial numbers of adults are crushed by cars each spring, on roads that divide upland sites and breeding ponds. Tunnels under the road can help solve this problem.

Herp highlight: Spotted salamanders may move more than 800 m (½ mi) from bodies of water where they breed, but usually will return to the same pond to breed year after year. Furthermore, individuals often use the same exact path each year from upland to aquatic sites, lumbering along at rates of 5–10 m per hour. It is obvious that knowing the behavior and natural history of these and other animals is very important when designing conservation plans. To adequately protect a population of spotted salamanders, it is necessary to protect both the aquatic site and the surrounding woods, perhaps up to distances of several hundred meters away from the water's edge.

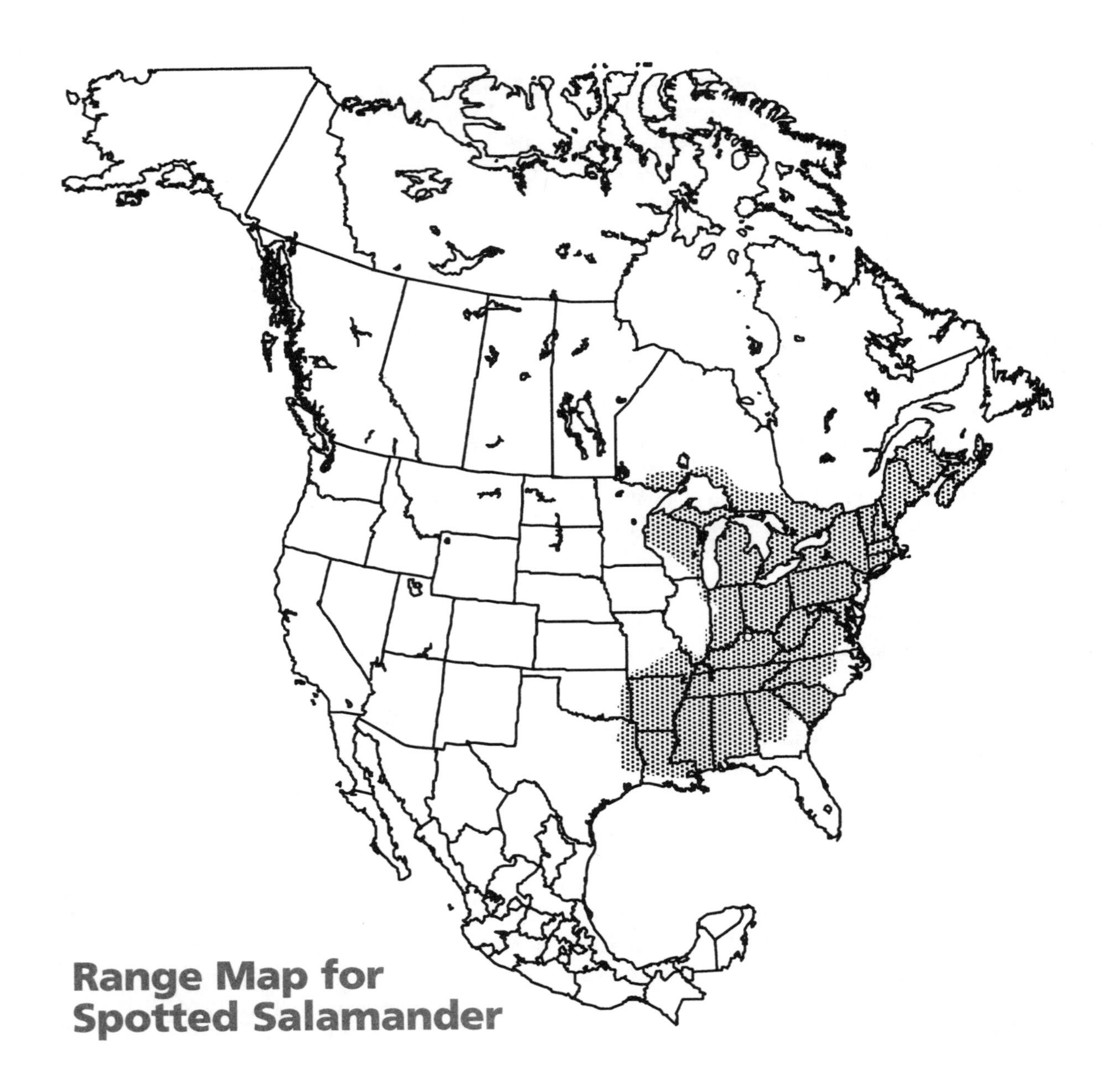

Range Map for Spotted Salamander

Species Name: American toad
Bufo americanus

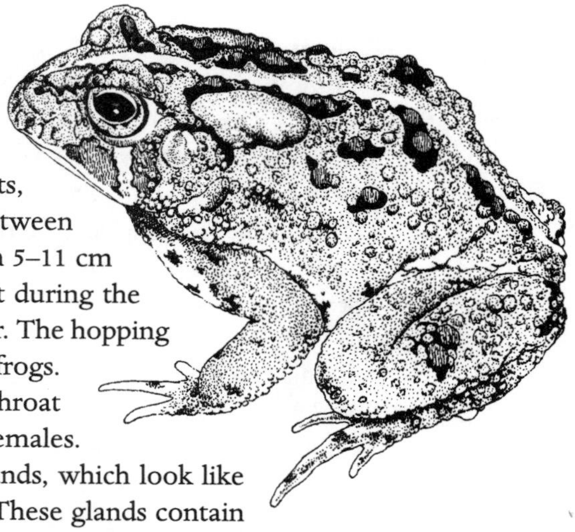

Description: The American toad is a chunky, brown, warty frog, with dry skin and black spots on its back, and a white belly with black markings. Each of the largest black spots on its back contains one or two large warts, and there are additional large warts on the rear legs between the knee and ankle. Adults, which usually measure from 5–11 cm (2–4 in), mainly are seen hopping about on land, except during the breeding season when they enter shallow bodies of water. The hopping of a toad is distinctly different from the leaping of other frogs. Males generally are smaller than females and have a dark throat and hardened, dark pads on their thumbs for grasping females. American toads and relatives have enlarged paratoid glands, which look like very large warts on the head, directly behind each eye. These glands contain toxic chemicals that cause illness or death in some mammals. Like all native frogs and toads, the American toad has a distinctive voice when calling for mates; its call is a pleasantly musical trill that lasts up to 30 seconds.

Similar species: In eastern Canada and northern regions of the northeastern United States, the American toad is the only hopping amphibian with dry, warty skin. In coastal and southern portions of the northeastern United States and along the Great Lakes states, this species overlaps considerably in its range with the Fowler's toad. This closely related toad usually has three or more smaller warts on each of the larger black spots, no enlarged warts on its lower rear legs, and few or no markings on its underside. It is hard to distinguish between these species when individuals are young. The American toad also overlaps on its southern boundary with the southern toad, which has raised crests on top of its head between the eyes.

Habitat: American toads can be found in almost any moist shady area on land, from the most remote country areas to yards in crowded cities. Their choice of aquatic breeding sites is equally broad, ranging from still areas of rivers, to ponds, ditches, ruts of roads, and even less-salty areas of salt marshes. During dry days, they usually remain under cover, slightly burrowed into the soil. They frequently hop around in the open at night and during rainy days.

Where and when: The American toad is widespread across the eastern half of the North American continent, occurring pretty much everywhere east of the Mississippi River, and extending impressively from the extreme northern areas of Labrador and Quebec to the Louisiana border. This species generally is not found on the coastal plain south of New England and along the Gulf Coast. In the northeastern United States, the activity season of American toads extends from mid-March through October. They are most active and moving about in the rain or on moist nights. Breeding adults begin to show up at bodies of water after the first warm rains, as early as mid-March; peak breeding occurs closer to mid-April. At this time, males can be heard and seen calling day and night. Occasionally, you may encounter a ball of males in the water, all entangled and holding tightly to each other, with a lone female somewhere in the mix.

Natural history: The American toad first reproduces at an age of two to three years. As temperatures begin to rise in the spring, adults migrate toward nearby sites with water. Often during warm spring evening rains, they migrate by the hundreds to their breeding sites. Males call to attract females, usually while sitting in shallow water or at the edge of the pond. The call of the male is a long, sustained, high trilling "bu-rr-r-r-r". During the breeding season, the toes of males become enlarged with dark rough skin (called nuptial pads); these are used to better grasp females during mating. Like most other North American toads, female American toads lay long strings of eggs (as opposed to other amphibians, which tend to lay eggs in clumps or individually). The eggs, often numbering from 2,000 to 10,000, usually are in two strings wrapped around vegetation or debris or sitting on the bottom. After about two weeks of peak breeding activity, adults have mostly moved back on land, leaving only their eggs behind. Eggs hatch very quickly, usually developing within 2 to 14 days. The larval tadpoles, small and black, often travel in large groups feeding on algae and plankton for 35 to 70 days, until they transform into miniature versions of adult toads and leave the water. On land, toads hunt and catch a variety of prey. They are considered to be beneficial to humans because they eat many things, such as insects, centipedes, and slugs, that are considered pests. They use a quick flick of their tongues to catch insects and other prey, and they use their front feet to stuff the bigger items into their mouths. They are voracious feeders and can easily be observed feeding, especially when they station themselves near outdoor lights and bug zappers on a summer night. In the North, between October and March, eastern American toads dig themselves deeply into the soil using their hind feet. There they go dormant through the winter months. Individual toads can survive for 10 years or longer in the wild.

Conservation status: American toads are abundant, widespread, and tolerant of a wide array of environments. They are an important component of many ecosystems, functioning both as predator and prey. The excretions of their paratoid glands are toxic and undoubtedly protect them from some predators. Nevertheless, some mammals like skunks and raccoons eat portions of toads, and many fish, snakes, and birds feed on entire toads. Thousands of American toads are killed each year while crossing roads during spring breeding migrations or as newly emerging juveniles. In some areas, warning signs along roads or tunnels underneath have reduced the impact of traffic on migrating amphibians.

Herp highlight: You will not get warts from touching toads. However, the paratoid glands, which are prominently displayed behind the toad's eyes, contain steroidal chemicals that can affect the blood and the heart of unwise predators. To avoid sickness, and even death, some clever mammals avoid the milky secretions and eat only the legs and the bellies of the toads. Some snakes, fish, and birds eat toads apparently without any ill effect. Humans can touch toads safely, but their secretions can be irritating to your eyes or mouth. Always be sure to wash your hands after handling toads and other animals.

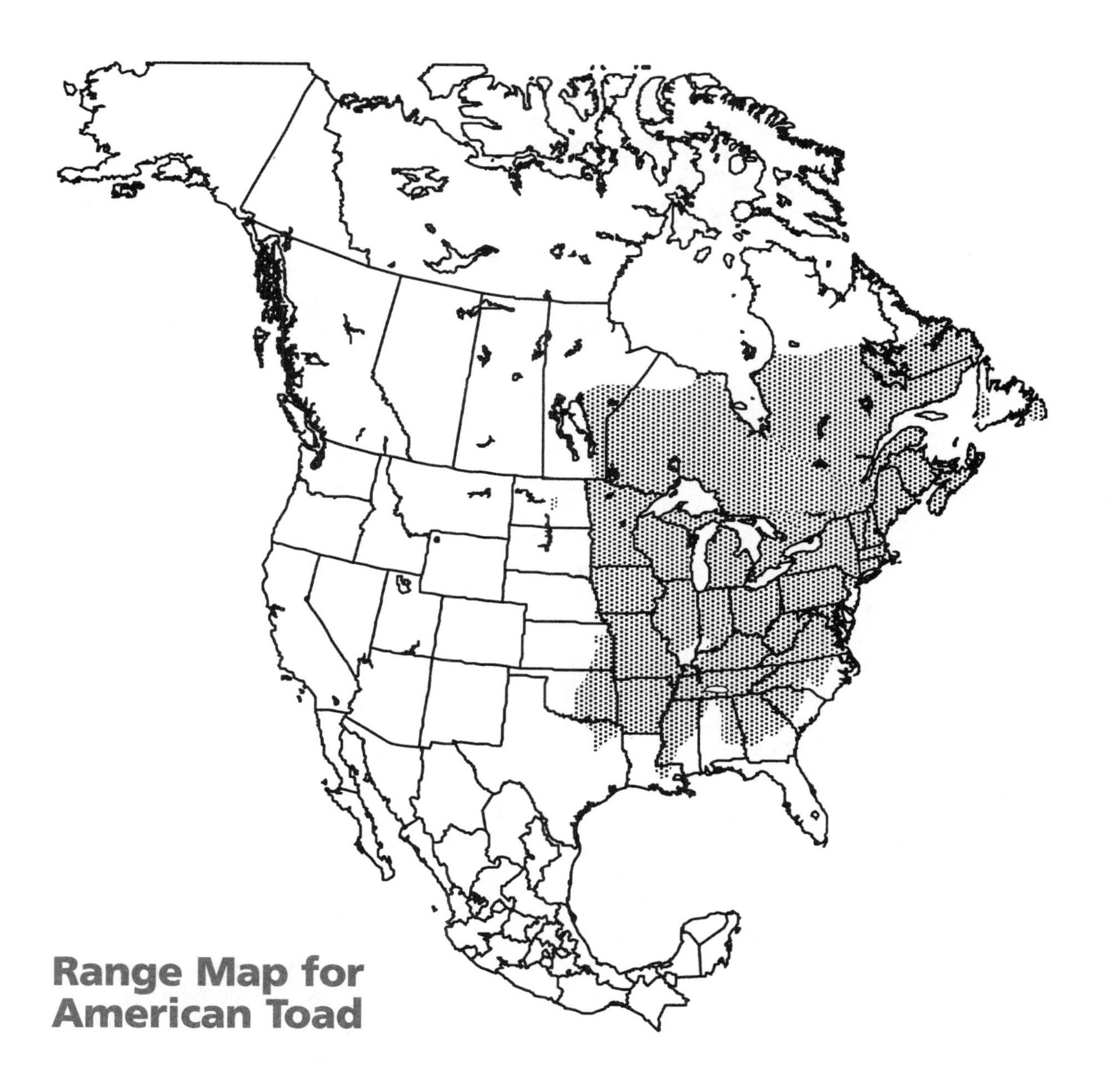

Range Map for American Toad

Species Name: Snapping turtle *Chelydra serpentina*

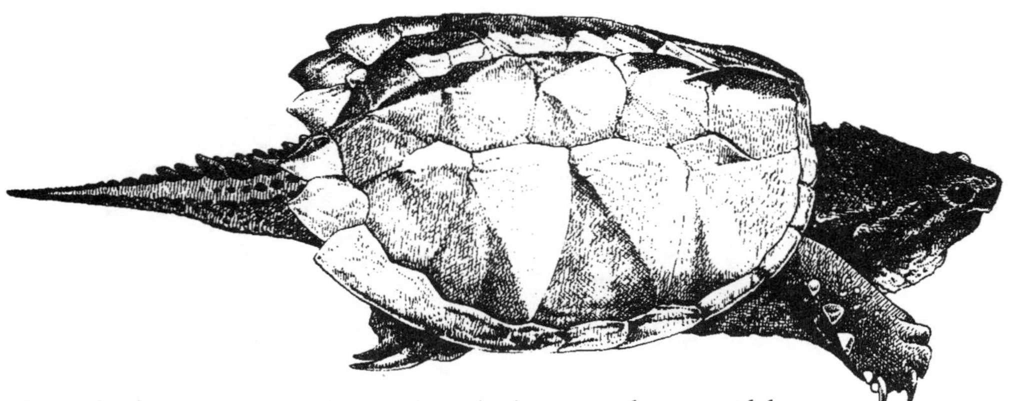

Description: The snapping turtle is the largest and most widely distributed freshwater turtle in North America. It has a long stegosaurus-like tail with a jagged upper surface, a stout head with a sharp hooked beak, an olive-green to black carapace (upper shell) that is jagged toward the tail end, and prominent claws on all four feet. These turtles can be pretty large, exceeding 35 cm (14 in) in straight-line carapace length, and weighing from 15–30 kg (33–66 lbs). On the underside, the plastron (lower shell) is yellow or grayish and quite narrow relative to other turtles, frequently giving the appearance that the turtle has outgrown its shell. The plastron is often darker in adults, ranging from light-brown to black. Although adult males are slightly larger than females, sexes can be difficult to distinguish without measuring plastron size and the distance to the anal opening. Juveniles that fall within the size range of other turtles are distinguished by their very long tails (as long as their carapaces), scutes with raised ridges on the carapace, and light spots at the edge of each marginal scute. Snapping turtles tend to remain partially submerged in the mud with only their eyes and nostrils protruding above the surface. In this position their heads resemble the head of a basking frog, except they are darker and more pointed. They are not often seen basking out of the water as are other aquatic turtles. Similar to most turtles, snappers do not tend to bite if stepped on underwater, nor do they bother swimmers. In fact, if you don't actually see a snapper, the chances are pretty good that you will never know it's there. The reason for their name is obvious, however, when they are encountered on land. Unlike all other turtles in eastern North America, they can be very aggressive, lunging their heads forward and biting with the slightest provocation (or sometimes just as a warning). With their sharp claws and fierce jaws, large snapping turtles can do considerable damage and are best left alone.

Similar species: All freshwater turtles in eastern North America have claws and webbed feet, but only snapping turtles have the jagged appearance to the rear of the carapace and the upper surface of the tail. In the south-central states, the alligator snapper can be confused with the snapping turtle, but the alligator snapper has a much larger head and lacks the jagged tail. Elsewhere, only young sea turtles are in the same size range as an adult snapping turtle, but they have flattened front flippers with one or no claws present. No other turtles are as threateningly aggressive on land as snappers.

Habitat: Snapping turtles can be found in any body of freshwater, small to large, from sea level to altitudes up to 2,000 m (6,560 ft) in North America. They occur in rural and urban areas, and even can be found in New York City's Central Park. Although some individuals enter coastal brackish waters, snapping turtles prefer slow-moving freshwater areas, with muddy bottoms and emergent vegetation.

Where and when: The overall distribution of the snapping turtle is very broad in North America, ranging from Nova Scotia to the Rockies in the North and from Florida to New Mexico in the South. Individuals in the northern region are active from March through October, and become mostly inactive during the winter. In the wild, snappers often are overlooked, as they spend much time buried in the mud of shallow waters. Although snappers in northern populations may occasionally bask out of water, they usually are only seen with their heads and sometimes upper carapaces visible at the surface. In contrast, they are highly visible during overland travel, which is commonplace among snapping turtles. Adults and juveniles can be seen moving about in early spring, while in early summer, nesting females frequently are seen walking, digging nests, or laying eggs. During the winter, they occasionally are seen moving slowly below the ice, but usually they remain dormant and burrowed in the pond bottom or along the banks.

Natural history: The snapping turtle is omnivorous, and will eat just about anything, live or dead. Its most frequent food items are aquatic plants and non-game fish, but it also eats insects, small mammals, young waterfowl, amphibians, and other reptiles. Snappers feed throughout the warmer months, but fast during the winter. Males can reach sexual maturity at the age of four or five years, while females may take several years longer. Average adult life spans of 20 to 30 years have been recorded in several studies, with some females living as long as 40 years. Adults can mate anytime from April to November, but the nesting period for females lasts around three weeks, from May through June, with a peak at the beginning of June. If a female mates after the nesting season, she will store the sperm until the following summer. Females prefer to lay their eggs on rainy afternoons and evenings, probably because the rain may help wash away scents that lead predators to the nests. Nesting females generally choose open, unshaded sites near wetlands, with well-drained sandy or loamy soils. They also are seen nesting on dirt roads and in soil supporting railroad tracks. The female first digs a nest chamber with her rear feet and claws, then fills the underground chamber with eggs. Eggs are spherical and pliable, like soft Ping-Pong balls that bounce and roll in the nest chamber. A single nest may contain from 20 to 40 eggs (even, if rarely, as many as 83). Successful eggs hatch from September through October. As with many other turtles, the length of incubation can vary by several weeks, depending on location and temperature.

Conservation status: Snapping turtles generally are abundant throughout their range, but in some areas are sparse due to several pressures. As in many other reptile species, snappers are highly vulnerable to predation at early life stages. Predation of nests in many areas is high, ranging from 30 percent to 100 percent of the nests in some studies. Main predators of the eggs, such as raccoons, crows, and dogs, are frequently associated with high human populations. Also, increased development often results in the loss of wetland and nesting habitat, which are both essential for snapping turtles.

Some isolated populations have been nearly wiped out by over-harvesting for their meat, and this decline is a major concern.

Herp highlight: The sex of snapping turtles and many other reptiles is determined by the temperature of the eggs while they are in the nest. Under warmer conditions (above 30° C) only female turtles are produced; at intermediate temperatures (from 24° to 29° C) males are produced; and in nests colder than that, females are produced. Interestingly, in some nests, the heat of the sun causes eggs in the upper nest to be warmer than eggs down deeper. This differential heating creates females near the top of the nest and males near the bottom. So, for sex determination, there is an element of luck involved in whether an egg was dropped into the nest early or late or, in some cases, the way in which the egg bounced and rolled as it fell. This environmentally controlled mechanism is called temperature-dependent sex determination.

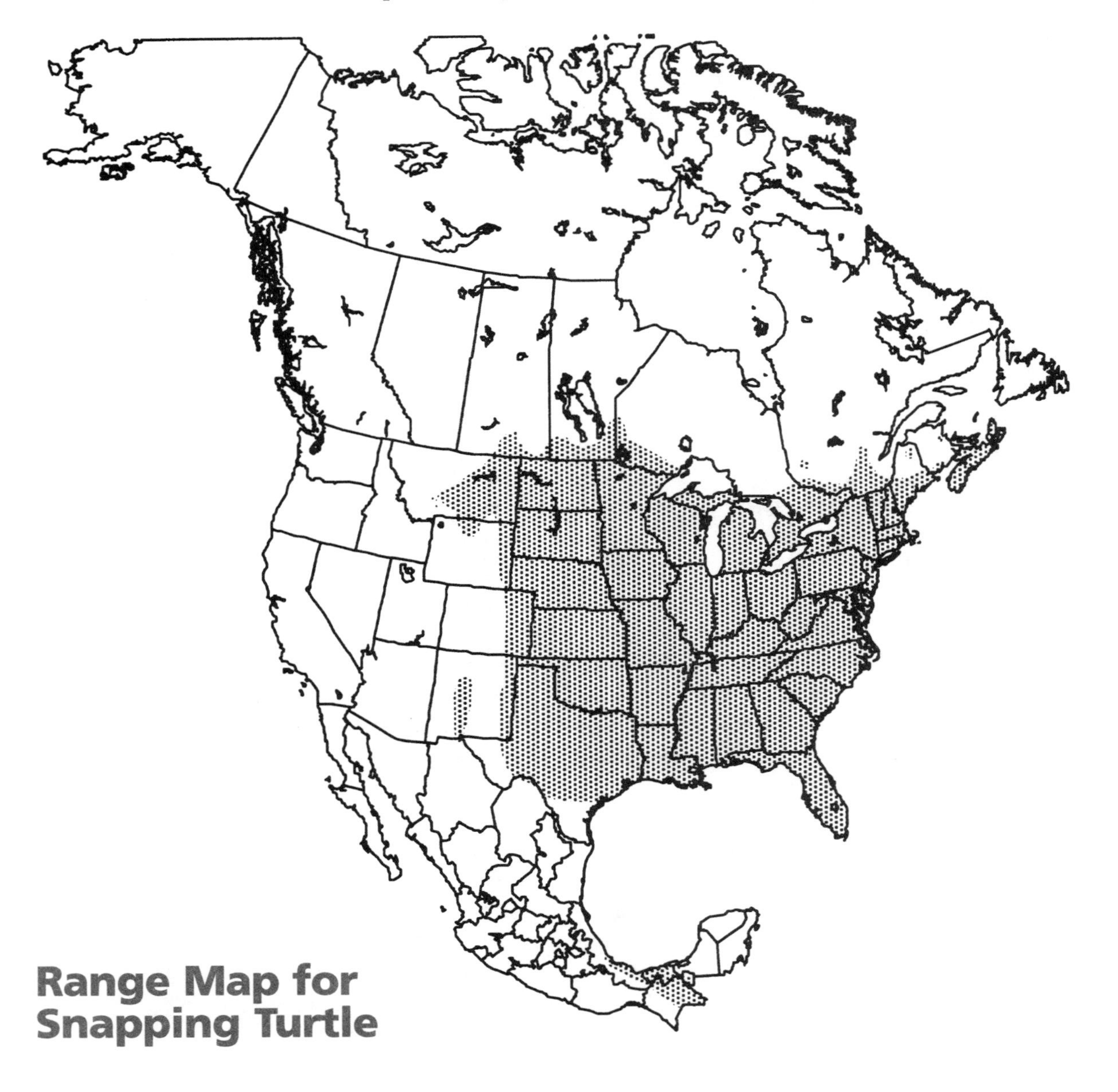

Range Map for Snapping Turtle

Species Name: Eastern newt *Notophthalmus viridescens*

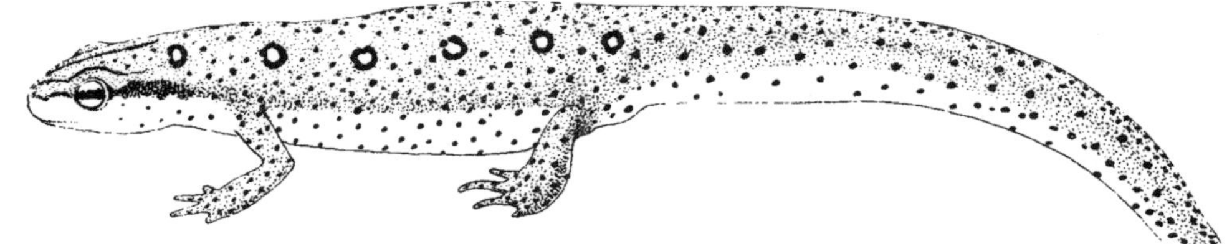

Description: The eastern newt is a small- to medium-sized salamander with two irregular rows of reddish spots bordered by black circles. The spots run down the length of the back and are present at all stages of the newt's life. However, the general appearance of the eastern newt is somewhat different in each of its three distinct life stages: larva, eft, and adult. As developing larvae in the water, newts are small, with faint red or yellow spots, bushy gills, and large tail fins. As they grow, they lose their gills, and in most populations, their tail fins disappears and they emerge from the water as brightly colored orange or red efts. Newts are the only type of salamander in the eastern United States and Canada that go through such an intermediate eft stage. During this stage, they are highly visible animals, frequently seen walking out in the open woods during the day. After several years of terrestrial existence as immature efts, they reach maturity and return to the water to breed with other adults. In many permanent ponds and lakes, they spend the remainder of their adulthood in the water. However, in some ponds, they move in during the breeding season and back out during winter or dry periods. In the adult stage, newts usually are dull brown, and have a yellow belly with numerous small black dots. Aquatic adults have flattened tails that are more appropriately shaped for swimming than the rounded tails of the efts. During the breeding season, the tail fin of the male gets very broad, and he often waves it around in the water, seemingly displaying his breeding status. Males also have a series of dark, hardened pads on the inside of their hind legs for clasping females. Adult newts usually range in size from 6–11 cm (2 ½ – 4 ½ in), while the brightly colored efts range from 3.8–9 cm (1 ½–3 ½ in).

Similar species: The red salamander, when young, often is bright orange. However, the red salamander has black dots, and similar to most other species of salamanders, has slimy skin and visible grooves along its sides between the front and rear legs. In the eastern region of North America, north of Georgia, the eastern newt is the only salamander that has red dots with black borders. In addition, very few other salamanders walk around in the open woods in broad daylight as boldly as do the bright orange newts in the eft stage.

Habitat: In both the adult and larval stages, eastern newts are aquatic animals that often live in great numbers in unpolluted, permanent bodies of water with plenty of aquatic plants. The species is extremely flexible, however, and can be found in temporary ponds, ditches, streams, and agricultural ponds. Eastern newts occur near sea level and along the entire eastern coastal plain; they are also numerous in many higher-altitude ponds and mountain lakes, occurring at elevations above 1,000 m (3,300 ft). The efts are found in a variety of terrestrial habitats, but mainly in moist woodlands that border the ponds where they originated. On damp days, hundreds of brightly colored efts and dull-colored adults may be out patrolling the forest floor.

Where and when: The eastern newt is common in eastern North America from Nova Scotia to Minnesota and southward to Florida and the Gulf Coast. This species occurs in coastal habitats, at higher elevation inland sites, and just about everywhere in between. Aquatic adults are active most of the year. In early spring, as the ice is melting in northern regions, they begin to congregate around the shorelines and around vegetation in preparation for breeding. They remain active throughout the summer and fall and in many areas can be seen swimming during the winter months. Occasionally, they will swim toward a hole cut into the ice of a frozen pond. In temporary ponds or in warmer regions, adults often go back on land during dry periods and throughout the winter, but return in very early spring to breed. Throughout the summer, larval newts can be seen swimming, until fall, when they transform into efts and move out of the water into the surrounding uplands. The bright orange terrestrial efts remain actively foraging until late fall, after which they settle under logs, in crevices, or in burrows until early spring.

Natural history: The eastern newt has a complex, and sometimes very long, life cycle. In the North, courtship and breeding take place in the water in early spring, usually from March through May. Some additional mating may occur in the fall, but females lay eggs only in spring. During mating, a male entices a female with a complex courtship dance, and embraces her for up to several hours with his rear legs. The female then follows along after him and picks up the packets of sperm that he deposits on the pond bottom. Once the sperm packets are brought inside her cloaca, the 200 to 400 eggs waiting there are fertilized. Over a period of up to several weeks, the female attaches these eggs singly to objects in the water. The eggs hatch around four weeks later, and the tadpoles develop into small newts over the next three months. They then emerge from the water and usually move to surrounding woods to begin a long existence as immature terrestrial salamanders, which are called efts. During this stage, which can last from two to seven years, efts are brightly colored and often are unconcealed, walking around in broad daylight. The bright color is an obvious warning to predators, meant to remind them that newts secrete toxic chemicals that make them distasteful or even harmful to eat. As the efts approach maturity several years later, their color becomes greenish-brown, their skin becomes smoother, their tails flatten out, and they return to the water to breed with other adult newts. On land, efts eat insects, worms, and other ground-dwelling animals small enough to swallow. In the water, the newts' diet includes mosquito larvae, aquatic insects, leeches, clams, and snails. The total life span of an eastern newt can be greater than ten years, and sometimes much longer.

Conservation status: The eastern newt is widespread throughout the eastern United States and southeastern Canada. In a wide variety of habitats, newts are an important component of both aquatic and terrestrial communities. As larvae, efts, and adults, they eat an impressive diversity of insects and other small creatures. They also are a food source for some predators, such as reptiles that apparently are not bothered by newt toxins. Because the terrestrial environment is so important to newts in many populations, their conservation depends on preservation of aquatic and surrounding upland habitats.

Herp highlight: Bright colors, such as red, orange, and yellow, often serve as a caution sign to potential predators, warning them of the toxicity of their prey. This vivid display of danger is called *aposematic* coloration. In the eastern newt, the bright orange skin of the eft contains glands that secrete a chemical that is offensive to other animals, often irritating their mouths, and potentially making them sick. The quickness of the negative reaction probably saves many newts, which may be left alone after the first bad taste. This helps explain why eastern newts walk around, seemingly without care, on the surface of the forest floor, boldly announcing their presence with flashy colors.

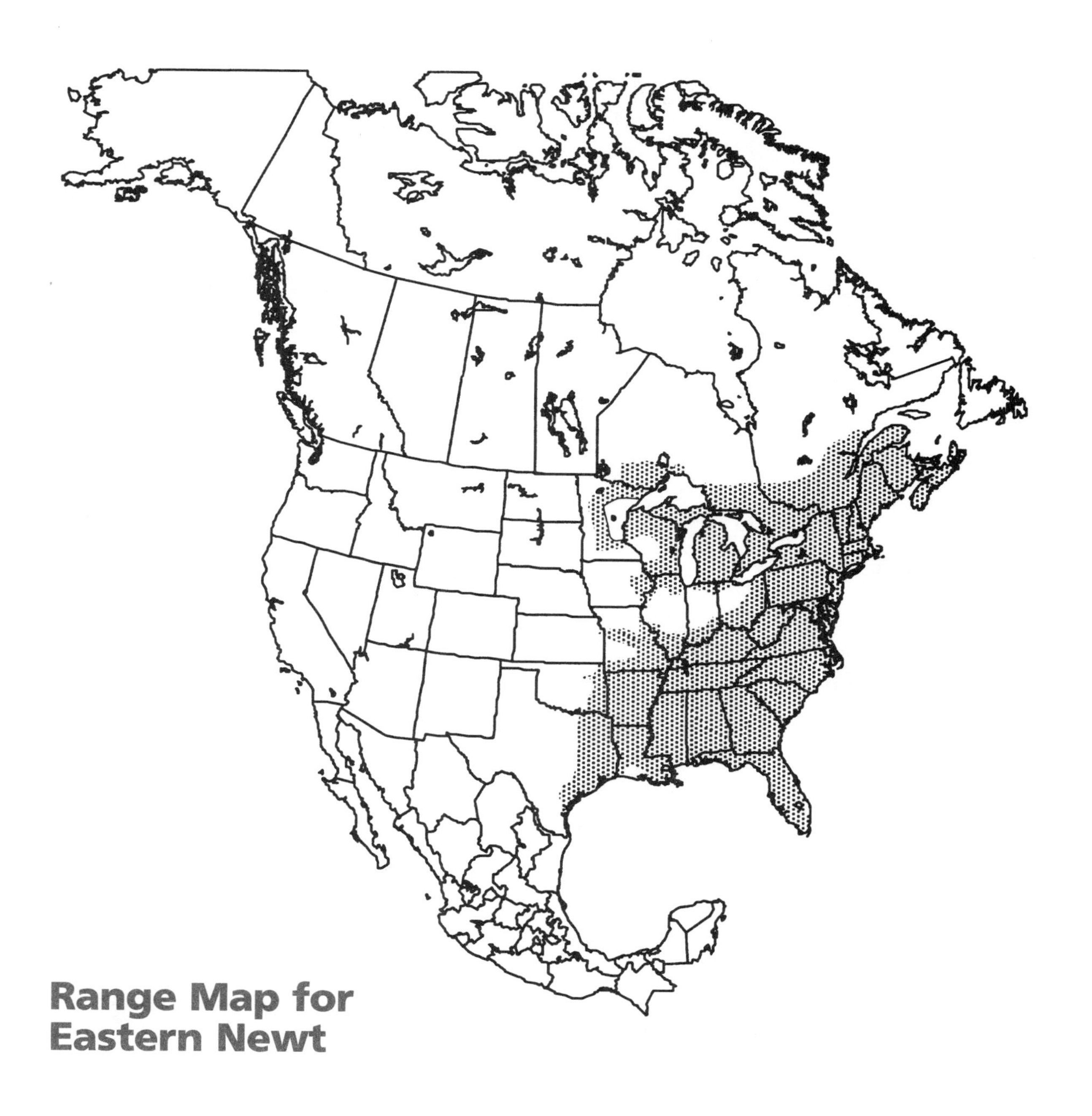

Range Map for Eastern Newt